LARVAL FORMS

And Other Zoological Verses

Walter Garstang

LARVAL FORMS

AND OTHER ZOOLOGICAL VERSES

WALTER GARSTANG

With an Introduction by

SIR ALISTER HARDY

and a Foreword by

MICHAEL LaBARBERA

THE UNIVERSITY OF CHICAGO PRESS

Chicago and London

Introduction and verses first published 1951 by Basil Blackwell; Presidential Address published 1929 in the *Journal of the Association for the Advancement of Science.*

The University of Chicago Press, Chicago 60637
The University of Chicago Press, Ltd., London

Printed in the United States of America

94 93 92 91 90 89 88 87 86 85 5 4 3 2 1

Library of Congress Cataloging-in-Publication Data

Garstang, Walter, 1868–1949.
Larval forms, and other zoological verses.

Bibliography: p.
1. Marine fauna—Larvae—Addresses, essays, lectures.
2. Evolution—Addresses, essays, lectures. 3. Larvae—Addresses, essays, lectures. I. Title.
QL121.G33 1985 592′.038 85–14114
ISBN 0–226–28423–9 (paper)

CONTENTS

Foreword *by Michael LaBarbera* vii

LARVAL FORMS AND OTHER ZOOLOGICAL VERSES

Introduction *by Sir Alister Hardy* 1
The Amphiblastula and the Origin of Sponges ... 22
The Invaginate Gastrula and the Planula 25
The Origin of Cnidoblasts and Cnidozoa 26
Conaria and Co. 29
Mülleria and the Ctenophore 30
The Onchosphere 33
The Trochophores 34
Mitraria's Fan Dance 35
The Ballad of the Veliger, or How the Gastropod got its Twist 36
Echinospira's Double Shell 38
The Nauplius and Protaspis 43
Kentrogon 44
Isopod Phylogeny 45
The Millipede's Egg-tooth 46
The Trilobites and After 48
Actinotrocha 51
Cyphonautes 51
Echinoderm Larvae and the Origin of Quinqueradial Symmetry 52
The Pentacrinule 54
Tornaria's Water-Works 55
Oikopleura, Jelly-builder 57
The Ancestry of Vertebrates 59
Leptocephalus brevirostris, the Larva of the Eel 60

The Axolotl and the Ammocoete 62
An Oceanographer's Dream 63
To a Herring Gull 65
Appendix: *A list of Walter Garstang's other Publications* ... 69

THE ORIGIN AND EVOLUTION OF LARVAL FORMS

1928 Presidential Address to the British Association for the Advancement of Science, Section D 77

FOREWORD

Larval Forms is a strange little book. Among invertebrate zoologists and embryologists it is notorious, for no other work conveys so well the complex mixture of thoughts and emotions the initiate feels on watching the graceful dance of an invertebrate larvum under the microscope. The lens displays only a delicate assemblage of cells swimming slowly through the water in loops and spirals, but to the biologist that sight is transformed by the knowledge that this small creature will, in a few days or weeks, undergo a metamorphosis as extreme as any caterpillar's or tadpole's into a creature of totally different form and habits; the knowledge that this creature is not merely a waystation between egg and juvenile, but a construct molded by natural selection to move, feed, and survive in a medium which, by the effects of size alone, behaves more like molasses than the fluid that fills bathtubs; the knowledge that this fairy-like tracery of cells has the two-fold mission of both finding an appropriate habitat for its post-metamorphic transformation and dispersing the species to new locales. Weighty stuff to be laid on a few micrograms of living tissue, but these tiny creatures have borne heavier burdens in the past.

The first two decades of the twentieth century, when Garstang penned the majority of these verses, were heady times in biology. The *concept* of evolution had long been accepted by the majority of biologists and paleontologists, but Darwin's *mechanism* of evolution, natural selection, was only then maturing. The rediscovery of Gregor Mendel's insights into heredity finally provided a genetic mechanism that would allow natural selection to work; the discovery of radioactivity at last lay to rest Lord Kelvin's contention that the Earth could not possibly be old enough for natural selection to have shaped the living world (a challenge made more severe by the fact that it came from the heart of quantitative science, physics). However, another and more subtle controversy—one initiated by a purported ally of Darwinism—had yet to be settled.

Ernst Haeckel was an early convert to Darwinism and probably the principal exponent of the concept of evolution in continental Europe. Haeckel, however, subtly shifted the focus of evolutionary biology from Darwin's mechanism of speciation and evolution to the reconstruction of ancestor-descendant relations—the deduction of phylogeny (a word Haeckel coined)—using embryology as his principal tool. To Haeckel, natural selection was only one component within the evolutionary process; for him, *the* law of evolution was his "biogenetic law": "Ontogeny is the short and rapid recapitulation of phylogeny . . ." and "[p]hylogenesis is the mechanical cause of ontogenesis. . . . The connection between [phylogeny and ontogeny] is not of an external or superficial, but a profound, intrinsic, and causal nature." As Stephen J. Gould points out in *Ontogeny and Phylogeny* (which should be consulted for more details on these controversies), in order for these statements to be true, adaptations *must* occur in the final stage of development (the adult): only then can a faithful "recapitulation" of phylogeny occur. It was this point that was the major perversion of Darwinism; Haeckel asserted that adaptation was restricted to the adult stages of organisms and dismissed larval or juvenile adaptations as rare aberrations.

Haeckel's popularity and influence were such that the assumptions of his biogenetic law became axioms for most of the influential biologists and paleontologists of the time. Using larval morphology as a literal record of phylogeny, Haeckel and his followers, for example, invented an ancestor for all metazoans (their Gastrea theory) and claimed to be able to trace the evolution of the gut from a minor dimple in the surface of the primordial metazoan (see page 25, below). Embryology was transformed from a study of the form and habits of early stages in the life history of organisms to a descriptive science whose sole goal was to reconstruct the form of the adult ancestors of living organisms from the morphologies exhibited during ontogeny. As Wilhelm His commented from the midst of the fray in 1888, "The present fashion requires that even the smallest and most indifferent inquiry must be dressed in phylogenetic costume, and whilst in former centuries authors professed to read in every natural detail some intention of the *creator mundi*, modern scientists have the aspiration to pick out from every occasional observation a fragment of the ancestral history of the living world."

Where does Walter Garstang fit into this story? From the beginning of his career, Garstang was intrigued by those features of larvae, particularly marine larvae, which were adaptations to *larval* life; Garstang appreciated that the morphology and behavior of larvae must be every bit as functional as the features of adults. Garstang championed the study of the functional biology of larvae and opposed the Haeckelian view (development as a literal record of phylogeny) as both sterile and inconsistent with modern concepts of natural selection and evolution. Since the turn of the century, proponents of the biogenetic law had been fighting a rearguard action; not only had Mendelian genetics removed any rationale for asserting that adaptations could only be tacked onto the ends of ontogenies, but the intellectual heart of recapitulationist doctrines, embryology, was increasingly turning to questions of ontogenetic *mechanisms*—the forces and transformations on the cellular level that produced the reorganizations and increased complexity seen during development.

In 1922, Garstang published a paper entitled "The theory of recapitulation: a critical re-statement of the biogenetic law" which sounded the death knell of literal readings of ontogeny as a record of phylogeny. After demonstrating that the biogenetic law was incompatible with Mendelian genetics and that the structure of larval forms had the capacity to canalize the course of evolution, Garstang, in phrases as trenchant as any of Haeckel's, pointed out that the evolution of new features in organisms must occur by the *alteration* of ontogenies: "Ontogeny is not a lengthening trail of dwarfed and outworn gerontic stages. Youth is perennially youth and not precocious age." "Ontogeny is not an animated cinema show of ancestral portraits; . . ." ". . . each ontogeny is a modification, within limits, of its predecessor; and by those predecessors the phyletic chain of adults was organized and equipped." "Ontogeny does not recapitulate Phylogeny: it creates it."

The verses presented in *Larval Forms* were, for the most part, written before 1922 and thus reflect these themes. Throughout you will find direct or indirect references to Haeckel and the biogenetic law, but here the refutations are light in tone if not in impact. You will find examples of the adaptations of larvae for larval life, a theme of Garstang's presidential address to the British Association for the Advancement of Science, also reprinted here. You will find allusions to the influence of larval morphology and adaptations on the evolution of entire classes and phyla of animals and the power of alterations to ontogeny (such as neoteny) to effect major transformations of form and habit in animals. The "hot topics" of modern evolutionary biology—life history strategies, macroevolution, heterochrony, developmental constraints—all appear in these verses; the jargon is missing, but the concepts are clear. Overlying all is Garstang's evident love for and wonder at these tiny marvels of creation. His verses were meant to be both tributes to larvae as organisms and a bit of light fun; although you may find some of these verses hard going because the protagonists are unfamiliar, I think you will conclude that Walter Garstang succeeded admirably.

Below are a few notes (keyed to page and line number) on some of the more obscure terms and allusions in Garstang's verses to help the non-specialist appreciate the subtle play of ideas. Armed with these notes, Sir Alister Hardy's comments (pp. 16–21) on individual verses in his introduction, and some background retained from a zoology course, the reader should be able to appreciate the wit of these verses. I have deliberately avoided explaining some obscure points since they have been explained in previous verses; the explanation of others arises in the course of the verse itself, and I wouldn't want to spoil the surprise.

The Amphiblastula and the Origin of the Sponges

(page 22)

10 **enteron** gut

11 **Monad** Haeckel's term for the original primitive unicellular organism.

(page 23)

1 **Monrad** A typographical error in the Blackwell edition; read Monad.

1 **seated lightly on your stalk** A group of living protozoans, the

choanoflagellates, forms colonies on a stalk jointly secreted by all the component cells.

7 **doctrines taught in Jena** Presumably a reference to Haeckel, but Goethe and Oken also taught there.

20 **the Collared Monad's tree** A delicious pun referring both to the stalk of choanoflagellates and Haeckel's penchant for representing phylogenies as (literal) trees.

The Invaginate Gastrula and the Planula

(page 25)

This verse is a critique of Haeckel's Gastrea theory; avoidance of the prose form does not decrease the potency of Garstang's venom.

The Origin of Cnidoblasts and Cnidozoa

(page 26)

cnidoblasts The stinging cells of sea anemones, jellyfish, and corals.

2 **diplothelial** Having only two layers of cells.

(page 27)

2 **myoplast** An epitheliomuscular cell; a cell in a coelenterate whose base is drawn out to form the equivalent of a muscle cell in higher invertebrates.

Conaria and Co.

(page 29)

1 **actinula** The larvae of *Velella*, the By-the-wind Sailor.

10 **scared them all to note** The "them" here refers to the fact that *Velella*, the Portuguese Man O'War and other siphonophores are not individuals but rather colonies of individuals, each highly specialized for a specific function such as feeding, propulsion, flotation, etc.

Mülleria and the Ctenophore

(page 30)

Muller's larvae The larvae of some marine flatworms.

(page 31)

4 **polyclad** One of the classes of flatworms.

The Trochophores

(page 34)

16 **nauplius** The first larval stage of most marine arthropods; again a backhanded reference to one of the recapitulationists' scenarios.

Mitraria's Fan Dance

(page 35)

19–24 Garstang refers here to the fact that the viscosity of sea water, the primary source of the drag force that slows the rate of the larvum's sinking, decreases drastically with a slight rise in temperature.

The Ballad of the Veliger or *How the Gastropod Got Its Twist*

(page 36)

The "twist" is the process of torsion during larval development in gastropods when the ventral portion of the larval snail suddenly and permanently rotates 180 degrees. The resulting morphological havoc such as crossed nerve cords, etc., seems not to bother the snails in the slightest. The explanation Garstang gives here for torsion, the character that *defines* the class Gastropoda, is still one of the two considered likely to be valid; more on this phenomenon can be found in the reprint of his presidential address to the BAAS that follows these verses.

7 **Archi-mollusk** The (hypothetical) ancestral mollusc.

(page 37)

9 **Lamarck** A reference to the inheritance of acquired characters; the first four lines on this page demonstrate that Garstang does not mean to be taken seriously about Lamarckian inheritance.

11 **stimulating stab** One of the classic demonstrations of comparative embryology is to poke a pre-torsion veliger with a glass needle; on retraction into the shell torsion suddenly and irrevocably occurs.

Echinospira's Double Shell

(page 38)

Echinospira The larvum of a prosobranch gastropod, *Lamellaria;* Garstang's 1922 paper referred to above gives a prose account of this creature and its biology.

4–8 The references to ammonites, "involute", "rostrated", etc., allude to Hyatt's classic study of "recapitulation" in the shell sculpture of lineages of the extinct cephalopod molluscs, the ammonites. Hyatt claimed that ammonites both ontogenetically and phylogenetically passed through a fixed succession of forms of shell sculpture: smooth—unituberculate—dituberculate—spiny—ribbed—smooth. The last character, although identical to the first, was supposed to represent racial senility.

(page 40)

10 **gerontic reminiscence** A reference to Hyatt's "racial senility."

The Millipede's Egg Tooth

(page 47)

3–5 Garstang refers here to the fact that, for small objects like larvae, the best way to increase drag (thus slowing sinking rate and increasing dispersal by ocean currents) is to greatly increase the animal's surface area with spines, setae, etc. The "salts were yet in very weak solution" refers to the then-current theories of Joly (1899), Sollas (1909), and Clarke (1910) that the oceans had originally been fresh water and had become progressively more salty with time. Our present knowledge indicates that the oceans have had approximately the same salinity since the Cambrian and within a factor of two of that value for the last three billion years.

8 **Proto-tracheate** Insects, millipedes, centipedes, and spiders have all (independently?) evolved a network of respiratory tubules extending throughout the body that are called trachea.

The Trilobites and After

(page 48)

2 **Paradoxus** A genus of trilobite.

5 **Cystid** An early crinoid (sea lily).

10 **variety is good** A reference to the Darwinian doctrine that variation within a species helps protect the species from extinction by providing a wider range of characters on which natural selection may act.

Actinotrocha

(page 51)

1 ***Phoronis*** A genus in the tiny (about 15 living species) lophophorate phylum Phoronida; their closest relatives are bryozoans and brachiopods.

14 ***Pro-cephalodiscus*** *Cephalodiscus* is one genus in another tiny phylum, the Hemichordata; the class Pterobranchia has many similarities to the lophophorate phyla.

Cyphonautes

(page 51)

A cyphonautes is the larval form of many of the marine Bryozoa.

Echinoderm Larvae and the Origin of Quinqueradial Symmetry

(page 53)

The last two lines of this verse eloquently express Garstang's view of the importance of the study of the functional morphology of larval forms.

The Pentacrinule

(page 54)

This story is retold in prose in Garstang's 1922 paper on recapitulation.

LARVAL FORMS

And Other Zoological Verses

NOTE

THE verses here published, although so light in character, were clearly intended by their author for the genuine student who is thoroughly familiar with the technical terms used in Zoology. The beginner in this science, however, if he is prepared to take the trouble and arm himself with a few text-books for reference, may well find these poems a most entertaining, if somewhat unorthodox, introduction to many important aspects of his subject. The volume must be almost unique: it shows us a scientist who had chosen the unusual medium of comic verse for the expression of some of his more profound and original contributions to biological thought.

A.C.H.

INTRODUCTION

THE late Professor Walter Garstang (1868–1949) was certainly a man of many parts.

To begin with he was well known for his pioneer work in pure marine biology: his delightful studies, for instance, of the natural history of sea-slugs and of burrowing crabs or on the adaptive significance of the colours of marine animals in general. He was on the staff of the Plymouth Laboratory on the day it opened in 1888, and he it was who started in 1895, the series of Easter classes which, together with their copies in other places, have become such a vital feature in modern Zoological education. All will recognise in him one of the founders of fishery research; he directed the English North Sea investigations from their beginning in 1901—with those classic early studies on the Plaice—to the time when he became Professor of Zoology at Leeds in 1908. On quite another side, many are those he has charmed with his poetical treatment of bird song[1] or entertained with *The Student's Opera*, his parody on Gay's immortal work. Apart from these activities, however, Garstang did something much more exciting; he added to zoological thought an idea which, I believe, will come to tbe classed among the more original and profound of those put forward in the last half century.

Why did so few of his contemporaries grasp the full significance of what he was driving at? Why even now, among the younger generation, are there not more who acknowledge the importance of the contribution he has made to the theory of evolution? There are, I think, three principal reasons. For two of them, it must be admitted, he was himself largely responsible. Firstly he was apt to let his passion for theoretical speculation run away with him; along with his more fully worked out ideas he would put forward, as if of equal value, others which many felt

[1] *Songs of the Birds*, 1922.

to be based upon much more slender evidence, and these consequently tended to prevent a more serious consideration of the former. Secondly there was his irrepressible sense of fun which caused him to present his views in lighthearted verse as much for his own amusement as that of his zoologist friends and pupils; of those who saw them thus decked out in comic dress most did not realise that Garstang thought them really important: they laughed at these merry jokes and rather naturally failed to take him seriously at all. Thirdly, his novel ideas came into direct conflict with what was then almost a zoological faith: the so-called Biogenetic Law of Haeckel or the theory of Recapitulation; was it to be wondered that his speculations—to many his airy flights of fancy—received but little attention in such an environment?

Garstang put forward his ideas in serious form in four important papers between 1921 and 1946; but they were not the easiest of papers to read and two of them were so specialized that many must have failed to realize their general evolutionary significance. It had always been his intention to publish his set of zoological verses; but always he delayed, hoping to add one or two more to them. Apart from several rhyming alphabets which were designed merely as mnemonics for his students, he left twenty-six of these little poems. They are published now not only because it was his wish that one day they should be made accessible—nor mainly because they must surely give pleasure and amusement to his many old admirers; they are presented as being what he himself intended them to be: the conveyors, in spite of their humorous style, of some of his more serious contributions to scientific thought. I certainly believe that he gets his ideas across with much greater felicity in these sparkling rhymes than he has done in all his more carefully calculated prose.

No one can pretend that they are all of equal merit as verses or of the same zoological significance. Some present original ideas of great importance, such as *Conaria and Co.*, *Mülleria and the Ctenophore* or *The Ballad of the Veliger;* others like *The Onchosphere* and *Kentrogon* are merely accounts of some of the more bizarre life-histories told in comic vein with the sole purpose of amusement. Nevertheless they are all included as he had originally intended and are arranged in the order in which he had placed them in a provisional contents table; the title of the collection, *Larval Forms,* is also of his choosing. Dates are given where possible, but only a few were dated exactly; some however are known to have been in existence in 1922 and are so indicated.

While the very brief comment on each poem, which will be given, may perhaps be helpful to some readers, the real intention of this introduction is to sketch in outline the nature of the fundamental contribution which I believe Garstang to have made to our ideas regarding the operation of the evolutionary process. The significance of a number of the pieces can only be appreciated if his main underlying conception is fully understood.

The development of his new idea sprang from his studies of the early stages in the life-histories of so many marine invertebrate animals which as adults live on the bottom of our shallow seas; these early stages were the *larval forms* which particularly interested him: those which drift freely for a time in the upper waters, as members of the plankton, before they finally settle down to be transformed into creatures of sessile habit. I vividly remember, from my undergraduate days, a remarkable lecture that Garstang gave in 1920 to the students of the Easter Class at Plymouth. In this he discussed the question as to whether these planktonic larval stages represent a primitive and ancestral adult type, as so many at that time thought, or do they represent special larval adaptations? He left one in no doubt that the second view was correct. With rapid sketches on the board he made a brilliant survey of the larvae of many different groups and showed how in each there is a compromise and adjustment between two rival needs—or, in other words, two competing selective advantages: on the one hand to grow up into the adult so as to reproduce the species, and on the other to remain floating as long as possible so as to distribute the species far and wide in the ocean currents—a particularly important matter for those forms with more sedentary adults. He showed clearly how the larval stages are as much adapted for dispersal as are the seeds or fruits of rooted plants; but he also showed how with all manner of different devices in different groups these two rival ' needs ' are met. He was one of the first to realize that selection will be acting just as powerfully upon the young stages in development as upon the adult end product; that in fact the developmental stages may become adapted to their particular mode of life just as much as the adult may. He realized that young stages, varying as much as older ones, may be modified in quite a different direction to that in which the adult structure is adapted—if the larva and adult inhabit two very different environmental fields. It is of course this great contrast between the adaptation of the young and old that has come about *pari passu* with the development—also by selection—of some marked form of metamorphosis in the life-history which alone has made the change from the one to the other possible.

Garstang realized, before anyone else, the profound influence which such larval adaptations might have on the evolution of the race. It was this that brought him into conflict with the many who still firmly believed in Haeckel's ill-named Biogenetic Law. Let us remind ourselves of the hold that this conception of recapitulation had upon zoologists at the time when he was developing his own ideas. How reasonable it then seemed that the animal during the course of its individual development from egg to maturity should be passing through, in broad outline, the stages it had passed through in the long history of the race in past aeons of time. What more likely than that the sequence of egg, blastula, gastrula, worm-like embryo and fish-like tadpole should reflect in concentrated form the evolution of the frog from its far distant protozoan ancestor? Could not development be looked upon as a kind of speeded up continuous modification of past ancestral adult forms? Imagine a photograph of each adult form in the same position, generation after generation, arranged in sequence of time from the beginnings of life to the present day, and imagine them joined one after the other like the consecutive shots in a cinema film, if such a film was now projected at speed, would it not give a representation of the growth of the modern adult from the egg? While they no doubt recognised that some modifications had been made to development in the course of time (caenogenic modifications as Haeckel himself termed them), the majority of zoologists thought of development as largely a recapitulation of past mature stages. Evolution was thought of primarily as a modification of *adult* forms. What else was worth considering since it was of course the adults that gave rise to each new generation? Were not such new modifications in the adult to be thought of as something added to the pre-existing form? Did not evolution consist in always adding something new to the adult end of development, and was not development gradually compressed, as it were, or speeded up, to make room for each new addition? Early in the century, when the knowledge of genes and their mutation was not what it is to-day, most zoologists, I believe, had some such faith. It is perhaps now worth reminding ourselves of a saying of that time: that nearly every scholarship boy who offered biology could be relied upon to write ' each animal in the course of its development climbs up its own ancestral tree ' or some phrase very like it. It was, of course, a conception dear to the heart of the neo-Lamarckians who, following Samuel Butler, regarded development as an epitome of age-long experience stored in the sub-conscious racial memory. MacBride's great *Text-book of Invertebrate Embryology* had only recently been published (during the first world war) when Garstang was preparing for his attack upon this facile view. Let me quote MacBride; he writes in his opening chapter:

'This theory is the so-called *fundamental law of biogenetics*, and is summed up in the phrase '*the individual in its development recapitulates the development of the race*.' If this 'law' can be substantiated the interest in embryology becomes immense, it binds all the innumerable phenomena of development into one coherent scheme, and opens the door to the hope that we may yet be able to sketch the main history of life on the earth.'

Finally MacBride sums up in his last chapter:

'The first and most far-reaching conclusion we may draw, is that, in general, the *larval phase* of development represents a former condition of the *adults* of the stock to which it belongs. This, in substance, is of course the *recapitulatory* theory of development, the famous *biogenetic law* of Haeckel.'

This whilst perhaps stating Haeckel's dictum in its most extreme form was not an isolated expression of its acceptance. It was still widely held when Wells, Huxley and Wells published their *Science of Life* in 1928: here is how they gave their views which, I think, reflect the more general opinion at that date:

'Exactly how far Haeckel's law takes us—what are its limitations: whether recapitulation ever shows us an animal's adult ancestors; whether present development ever does more than recapitulate ancestral development; what is the cause of recapitulation and why some of the characters and structures are regularly recapitulated in development, others only occasionally or not at all—all this we cannot here discuss. What we are here concerned with are the positive, visible facts. Tens of thousands of animals do recapitulate the past during their development—do, without any apparent advantage in so doing, show organs and constructions which occur elsewhere in the adults of less specialized creatures; and in none of these tens of thousands of cases is this departure intelligible save on the view that in so doing they are repeating phases that were once final forms in the earlier evolution of the race.'

Garstang from his study of larval forms realized that such types as trochophores, veligers, plutei and a host of other kinds were not likely to be past adult forms ' pressed back into early life,' but young stages specially adapted for planktonic distribution. He opened his attack upon Haeckel's and MacBride's views in his paper. ' The Theory of Recapitulation: a critical Restatement of the Biogenetic Law,' which he read to the Linnean Society of London in 1921.[1] Here he called modern zoologists back to the almost forgotten ideas of Meckel and von Baer of a century before. It was von Baer's laws which expressed the view that the life-histories of animals along their whole lengths tend to deviate more and more from the course of development which was followed by a distant ancestor. They were laws which clearly indicated, for example, that the transitory gill pouches of a mammal are not to be thought of as the equivalent of those of a once ancestral *adult* fish, but rather those of a corresponding *embryonic phase* of such a far off relative. In his paper Garstang used for the first time the term *Paedomorphosis;* in summing up his views he says: ' Ontogeny does not recapitulate Phylogeny: it creates it.'

It is in this term—Paedomorphosis—that he crystallised the great new idea that came to him: the idea that the adaptations to a larval or young life might have not only a profound influence upon the adults of the race, but upon the whole course of the future evolution of the stock concerned. The importance of the idea was only dimly fore-shadowed in his paper of 1921. It was not until six years later that he gave two brilliant examples of different ways in which his principle could act.

His first example of paedomorphosis was given in his Presi-

[1] *Linn. Soc. Jour. Zool.*, xxxv. pp. 81–101, 1922.

dential address to the Zoology Section of the British Association at its meeting in Glasgow in 1928; this he devoted to larval adaptations with special reference to the Mollusca. He explained the characteristic ' torsion ' of the gastropod body—a twisting of the viscera through 180° quite distinct from the independent spiral of the shell—as the result of a developmental adaptation to bring about a greater protection for the more vulnerable head region of the pelagic larva. This idea is beautifully presented in the *Ballad of the Veliger or How the Gastropod Got its Twist* which, except for the ode *To a Herring Gull* which is different from the rest, is the only one of these zoological poems to be previously published; it was privately printed and circulated at the same time as his British Association address. I think most will agree that this little poem did more to impress this idea on the minds of zoologists than the more prosaic prose of the formal address. Here was one way in which paedomorphosis might operate—selection in favour of a particular larval adaptation might be so strong that it could outweigh any contrary selection in adult life and so impose the larval character upon all the adults of the race. Interesting as such an idea is it is much less profound than the second way in which Garstang showed that larval adaptations might affect the course of evolution.

The second and much more significant example of paedomorphosis he gave in his great paper on ' The Morphology of the Tunicata, and its bearing on the Phylogeny of the chordata '[1] which he published in the same year as his British Association address (1928). In this, besides a host of points bearing on the evolution of the tunicates and lower chordates, he put forward his view that the chordates were derived by paedomorphosis from the pelagic larval forms of a *sedentary* invertebrate animal, and that the modern ascidians are *not* degenerate chordates, as was usually held, but are animals living much as the original ancestors

[1] *Quarterly Journal of Microscopical Science*, Vol. 72, pp. 51–187.

of the chordates once had lived, sending up their planktonic larvae to spread the species far and wide. Here he coupled the idea of paedomorphosis with that of neoteny. We have previously noted his early realization of the rival selective forces upon pelagic larvae (*a*) that favouring long floating powers for maximum distribution and (*b*) that favouring the more rapid onset of maturity to lead to more prolific reproduction. The latter effect may be brought about by the acceleration of the delopment of the reproductive organs in relation to that of the rest of the body; this *is* in fact the phenomenon of neoteny.

Perhaps a brief digression to give a more familiar example of neoteny may be desirable to make Garstang's ideas clearer. The Mexican Salamander in its fully developed condition is the *Amblystoma* which has lost its gills, breathes by lungs and spends most of its time on land; more frequently this same species becomes sexually mature and breeds in a condition of body which is really larval: this is the *Axolotl*, which keeps its external gills and gill slits and remains in the water. In the Axolotl the development of the reproductive organs has been accelerated in relation to the rest of the body so that it is neotenous; but here we see a species which is only partially so, for it may at times be converted, either in nature or experiment, into the fully adult *Amblystoma*. Some other newt-like forms such as *Necturus* and *Proteus* are permanently in the neotenous condition; here we see an evolutionary change brought about by the fixing of this change in the rate of development of the gonads. It has been known through the discovery in 1927, by Ford and Huxley working on the development of the eye colouration of the amphipod crustacean *Gammarus*, that mendelian genes can affect the *rates* of different developmental processes: accelerating or retarding some in relation to others; this explains the mechanism of neoteny.

Let us now return to Garstang's view on the origin of the chordates or vertebrate stock. He imagined a sessile invertebrate animal living like an Ascidian (a sea-squirt) and sending up planktonic larvae. He then imagined the dual forces of selection acting upon this larval form to such an extent that (a) its powers of remaining afloat were prolonged by the production of more and more motile forms having a muscular tadpole-like tail and (b) the development of its gonads was greatly accelerated by neotony. Finally he imagined this prolonged larval stage becoming actually sexually mature while still in the plankton: just as the Axolotl may become sexually mature while still in the aquatic phase. Now the more dramatic step is easily taken —the former sessile bottom-living adult stage can be eliminated, as indeed the once lung-breathing former adult Necturus was surely eliminated. Evolution takes the remarkable step of leaping from an apparently lowly sedentary form to an active free-swimming type—the forerunner of a new race to lead on to the vertebrates—to fish—to Man! Now were the tables turned with a vengeance upon those who clung to the theory of recapitulation; instead of past adult forms determining the course of development of later members of the race, we see proclaimed the idea of developmental novelties changing a major line of evolution! No wonder that some of the older zoologists, more set in their ways of thought, regarded this as a monstrous piece of speculation. Speculation, no doubt, it is—but no more so than the idea of recapitulation which it overthrows—and moreover it is an idea based upon a much more substantial array of facts.

The ascidians, or tunicates, actually do send up little tadpole-like larval forms which keep themselves up in the water by the development of muscular undulating tails rather than relying upon ciliated bands as do most of the other planktonic larvae. Now there is one group of the Tunicata, the Larvacea, which has every appearance of having done just what Garstang supposed the ancestors of the vertebrates to have done; they are pelagic adult forms and there can be no shadow of doubt that they are closely related to the more usual sessile forms; there can be little doubt that they are in fact permanently neotenous one time larval forms now specialised for their particular mode of life. Garstang did not imagine that the main chordate stock was actually derived direct from an ancestor like a modern ascidian—but he believed the ascidians and their tadpole-like larvae gave the clue. As early as 1894 he had made a remarkable comparison between the early chordate embryo and the larvae of the echinoderms. He had not at that early date suggested the origin of one from the other by paedomorphosis—he had so far only stressed their relationship. Now in 1928 he suggested that the chordate stock had arisen from some sedentary ascidian-like invertebrate of the distant past having a common origin with the echinoderms. So important is this contribution to zoological theory that I shall emphasise it by quoting an excellent appreciation of it in Dr. de Beer's book *Embryos and Ancestors* (pp. 52–54):

'Garstang was the first to look for the trace of the ancestors of the vertebrates in early instead of adult stages of invertebrates; and he focused his attention on the larvae of Echinoderms (starfish, sea-urchins, sea-cucumbers, etc.). He showed that if the ciliated bands on the larva (auricularia) of a sea-cucumber were to become accentuated and rise up as ridges leaving a groove between them, and if these ridges were to fuse, converting the groove into a tube, a structure would be produced which has all the relations of the vertebrate nervous system, including such details as the neurenteric canal. Not only this, but the two modifications of the vertebrate nervous system which are found in Amphioxus and all higher forms on the one hand, and in sea-squirts on the other, can be based on differences which are found in the disposition of the ciliated bands on different kinds of Echinoderm larvae. This theory of the origin of the vertebrate nervous system has several advantages. In the first place it avoids the difficulties which beset any attempt to derive it from the existing nervous systems of any other invertebrate. It also agrees with the principle of neurobiotaxis, according to which a concentration of nervous tissue takes place in the region of greatest stimulation. If the ancestors of the vertebrates had crawled about on their ventral surfaces like most invertebrates, one would expect their nervous system to be ventral, like that of most of the invertebrates. But the nervous system of vertebrates is dorsal, and it is precisely the dorsal side of the body which would receive the greatest stimulation in a form swimming freely in the sea, the stimuli being the rays of light penetrating through from the surface. Not only is the auricularia free-swimming, but it bears an unmistakable resemblance to the tornaria larva of *Balanoglossus*, and *Balanoglossus* is an undoubted relative of *Amphioxus* and the early vertebrates (or chordates as they may more correctly be termed). But the resemblance between the Echinoderm larva and the chordate goes further still, for the former has an adoral ciliated band formed partly from the inner layer of the body; and in a corresponding position the chordates have a ciliated band called the endostyle, which is looped in the same peculiar manner as the other. The middle layer of the body arises in three tiers or segments in the Echinoderm larva and in *Balanoglossus*, and indications of the tripartite arrangement are present in *Amphioxus*, Further, the body-cavity of the Echinoderm larva is in communication with the outside by a pore, as in *Balanoglossus*, *Amphioxus*, and several other chordates. In fact, if the nervous system and endostyle are formed in the way suggested, all that is required to turn the Echinoderm larva into a chordate is the formation of the notochord and the piercing of the gill-slits.'

More than anyone else de Beer has recognized and acknowledged the importance of Garstang's conception of paedomorphosis and has himself coined another term to stand beside it—

gerontomorphosis—to signify evolutionary changes brought about by modification of adult structures. His brilliant book *Embryos and Ancestors* (called *Embryology and Evolution* in its original form) is largely a discussion and analysis of those two factors in the process of evolution. To paedomorphosis he assigns the larger steps in the course of evolutionary change and to gerontomorphosis the smaller ones. He goes on to show how the larval forms of a Myriapod, soon after hatching bears a great resemblance to an insect: how walking limbs are first developed upon the three segments immediately following the head, just as on the insect thorax, while the remaining segments bear only rudimentary ones as do the abdominal segments of some of the most primitive living insects of today. To see the larval myriapod is to be at once reminded of an insect; and since the evolution of the insects from myriapod-like forms is supported by so many other considerations it is not so wild a speculation to suppose that the insects like the vertebrates, have arisen by a process of paedomorphosis involving neoteny. Bolk (1925) had shown that man himself was probably the product of a neotenous transformation taking place in the anthropoid stock; he termed the process *feotalization*, whereby embryonic characters of the past have become incorporated in the adults of the present.

Garstang's next major contribution to the development of the idea of paedomorphosis came in his large paper on ' The Morphology and Relations of the Siphonophora ' published in 1946,[1] only two years before his eightieth birthday. For more than a century two opposing views as to the nature of the siphonophores have had many advocates: the ' poly-organ or medusome hypothesis ' which supposes them to be highly developed medusae complicated by secondary budding, and the ' poly-person or swimming hydroid colony hypothesis '. Garstang reviews the evidence for each and makes a detailed study of the different

[1] *Quarterly Journal of Microscopical Science*, Vol. 87, pp. 103–193.

accounts of siphonophore morphology and bionomics; he did not favour either view, but came to a new and original interpretation which might perhaps be called the 'actinula hypothesis'. He believed that the siphonophore colony had arisen by paedomorphosis with the precocious development of a floating larval polyp form sent out from some original sessile hydroid stock: indeed a larval form comparable with the actinula larva of *Tubularia.* This we shall see hinted at in his little rhyme *Conaria and Co.*

Earlier he and his brother-in-law the late Dr. Robert Gurney in a joint essay on the evolution of the Crustacea from Trilobite ancestors made passing reference to their view that the Cladocera are paedomorphic in origin. Gurney himself later produced careful evidence to suggest that the Copepoda—one of the most successful of all crustacean groups—are similarly derived from the pelagic protozoeal larvae of bottom-living decapod-like forms. This view has recently been commented on favourably by no less an authority on the crustacea than Dr. W. T. Calman.[1] Only a few months before he died Garstang had drafted a communication to *Nature* to put forward his latest suggestion that *Amphioxus* might be regarded as a paedomorphic ammocoete-like larva of a Cyclostome; it was never sent, because the day on which he was to have posted it he found that the whole of his idea had recently and quite independently been published by the great Stensio.

Some of his other paedomorphic ideas, for example the evolution of the ctenophores from the planktonic eight-lobed Müller's larva of a bottom-living polyclad, are set out in his verses.

Speculation and still more speculation! Many may ask, what can be of real value in such exercises? It was, as I have already hinted, his love of these conjectures which prevented more of his

[1] Letter quoted on p. 119, *Proc. Linn. Soc.*, Vol. 162, 1950.

contemporaries from taking a serious interest in his views; had he lived a generation earlier, he would indeed have been a leader of zoological thought. He was a student at Oxford in the 'eighties'—in that heydey of speculation as to the origin of the different groups of animals. Then followed his period of active marine work. Twenty years later he returned to a more academic life as Professor and took up once more the problems of phylogeny which had fascinated him in his undergraduate days; in the meantime however the main trends of zoological interest had flowed into channels of a more experimental nature and he found that his theorising was thought old fashioned and out of date. 'Just another amusing Garstang speculation' his friends said as they turned back to measure the alkalinity of this or the oxygen consumption of that particular animal they were interested in.

Of course none of Garstang's speculations can ever be established by experiment—they can never be proved as can an hypothesis in physiology. Many of them may be entirely wide of the mark. But the point I want to make is one which is independent of the fact that any particular one of them is true or false. It is this: Garstang with his concept of paedomorphosis has altered our whole outlook on the process of evolution.

In an imaginary wide view of the streams of life we see the various evolutionary lines branching again and again as they advance in time; branch after branch tends to become so well adapted to some particular way of existence that it becomes eventually highly specialised. Such older specialised lines have been thought to be doomed to eventual extinction as conditions must in the long run change; the lines of new development have been thought always to spring from the less specialised forms. This indeed must often be so. But is it the *inevitable* rule it was thought to have been only a little while ago? Some have taken so extreme a view as to suggest that there could be no new major lines of evolutionary change in the future, because all the different lines of animal organization have now become too specialised to give rise to any other new kind! Garstang's speculative thoughts have enabled us to see that specialisation need not always be the end of evolutionary progress. However specialised the *adults* of a stock may have become, it is still possible for their younger stages to become modified in other ways and then by neoteny to produce a new paedomorphic line to blossom forth to become—who knows?—perhaps a whole new order—class—or even phylum.

Siphonophores, ctenophores, cladocerans, copepods, insects and the very vertebrates, including Man himself, have all—and others too—been shown possibly, if never certainly, to have had a paedomorphic origin each from something very different. This is the major contribution that Garstang made to zoological thought: to have first shown that such a possibility is not unreasonable. I, for one, am confident that the coming generations of zoologists will judge it to be among the more fundamental conceptions given to our science in this century.

Let us now see the point being made, as it has never been made in public before—not by the late emeritus Professor in cap and gown, but by the impish boy in him, in the garb he much preferred: the cap and bells!

His verses will speak for themselves. It only remains to make the briefest of comments and explanations. Many, indeed most, as has already been pointed out in a foregoing note, will only be understood by the zoologist who is thoroughly familiar with the technical terminology of his science. Any attempt to interpret all the terms to the general reader would be quite impossible; it would require a little biological dictionary rather than a glossary. They are clearly intended solely for true students of the subject. Each little poem dealing with a different group of animals, is arranged according to a usual classificatory sequence which starts with the simpler invertebrates.

It is perhaps a pity that *The Amphiblastula and the Origin of*

Sponges must come first, for it is I think not one of his best; it cannot help being somewhat involved like the different types of sponge development and evolution he is describing. *The Invaginate Gastrula and the Planula* is a cunning hit at one of the recapitulationists most cherished dreams. *The Origin of Cnidoblasts and Cnidozoa* shows Garstang in his most speculative mood; it summarises a theme he developed in the last paper he read to the Linnean Society shortly before his death: a paper which, however, was never published. Few I think will follow him in this, which, I must confess, appears a most extravagant flight of fancy; but time perhaps—who knows?—may yet show that I, and some others at least, are too fixed in our grooves of zoologic thought!

Conaria and Co., in the most light-hearted vein gives a thumb-nail sketch of the development of that remarkable composite and surface-living animal *Velella* from its deep-water larva, Conaria; in it he just gives a hint of one part of his great 'Actinula hypothesis' concerning the origin of the Siphonophora. The next, *Mülleria and the Ctenophore*, tells in charming pantomime, via an imaginary conversation between the two, of the possible origin of the latter by a paedomorphic and neotenous modification of the former: 'Müller's larva' an eight-lobed planktonic stage in the development of a polyclad turbellarian—a bottom-living flat worm. *The Onchosphere* is for sheer amusement and presents the enormities of a tapeworm's life and development in terms of knock-about fun:

> 'But if a pig should swallow him, as many porkers do,
> He's made a start with no mistake: he's on the road to you!'

In *The Trochophores* and *Mitraria's Fan Dance* he returns to the description of pelagic larvae and deals with the various more interesting adaptations in the development of the polychaet worms; in the former is just a hint of his views as to the origin of the Crustacean *nauplius*. Then comes *The Ballad of the Veliger*; to my mind the happiest and gayest of them all, and one giving a most graphic expression of one of his more important contributions to zoological theory in his explanation of how the gastropod got its twist. Its cheerful mixture of nautical and marine biological language gives a feeling of the very sea itself, and his powers of description bring the tiny creature, with its two little locomotive ciliated lobes, very much to life:

> ' The Veliger's a lively tar, the liveliest afloat,
> A whirling wheel on either side propels his little boat;'

Following this frolic comes a somewhat long and detailed poem intended rather for the specialist: *Echinospira's Double Shell*. In it he describes his own observations and those of his friend, Dr. Marie Lebour, on the development of this curious little double-shelled young phase—the *Echinospira*—of the gastropod mollusc Lamellaria. It is a poem that will well repay the most careful reading by the student; in this unusual medium he actually presents his own, hitherto unpublished, observations through his microscope of the very remarkable use of the larval operculum as a tool modifying the shell as growth proceeds. This must be something almost unique in the history of scientific communications.

In *The Nauplius and Protaspis* he gives a sketch of his views on relations of the crustacean nauplius larva to that of the extinct trilobites—and further the relations of that to the Annelid *Nectochaeta* larva more fully referred to in the earlier poem *The Trochophores*.

Kentrogon, in the same comic vein as that of *The Onchosphere*, describes the truly amazing phase of the life-history of the parasitic barnacle *Sacculina* where the little *Cypris* larva settles on its victim, a crab. He describes how this gives rise to another stage called the kentrogon, which is little more than a living hyperdermic syringe evolved for the purpose of injecting its most vital contents into the body of its new host.

> 'Through the antenna of the *Cypris* on the hairlet of the crab
> His fine proboscis travels and inflicts the needful stab:
> Then, gathering all the salvage that he's rescued from the rout,
> He slips along the tunnel of his own projecting snout.'

Is it not possible that a student may find that a more easily memorized—if not more memorable—account of this most extraordinary event than one in prose? Yes, I'm sure it well deserves the light of publication.

Now comes one of the gems of the collection: *Isopod Phylogeny*, a deadly satirical shaft—dressed up as nursery rhyme—directed at the recapitulationists and his old adversary in particular:

> 'Sing a song of six legs, a new phyletic stage!
> Four and twenty Isopods cradled in a cage:
> When the cage was opened, out they ran to play,—
> Wasn't it a jolly thing to have a jolly day!'

Then follows a most delightful hit and a piece of sound zoological observation too—but I must not spoil the fun to come, for . . .

> 'MacBride was in his garden settling pedigrees . . .'

The Millipede's Egg-tooth is frankly a piece of Garstangian speculation attempting to derive the dorsal spine with which a larval millipede breaks out of its shell from similarly placed dorsal spines on the larvae of the long extinct trilobites. In *The Trilobites and After* he gives his views as to how that remarkable fossil crustacean *Lepidocaris*, discovered by the late Dr. Scourfield, must have been evolved from the earlier trilobites; the argument is presented in the form of a dialogue between the *Lepidocaris* in the water and *Arthropleura*, one of the most primitive land arthropods on the bank nearby. He then returns to planktonic larval forms and in two little rhymes, *Actinotrocha* and *Cyphonautes*, briefly suggests possible relationships from a consideration of these larvae of *Phoronis* and the ectoproctous Polyzoa respectively.

In *Echinoderm Larvae* he sketches the main differences in the larvae of the different classes of echinoderm and speculates upon a possible *larval origin* of that quinqueradiate symmetry of the group. The recapitulationists are again attacked in *The Pentacrinule* in which he has an imaginary conversation with this young stage of the crinoid feather-star *Antedon*, showing how many of its features are developmental adaptations peculiar to itself instead of being repetitions of ancestral traits.

We come at last to poems of the Chordata. In *Tornaria's Waterworks* he describes the architectural beauty and wonderful mechanism of the pelagic larva of the hemichordate Balanoglossus; and in *Oikopleura Jelly-builder* he gives a beautiful account of this extraordinary creature's intricate apparatus which it makes out of jelly for sieving the smallest of living specks from the water for food.[1] The *Ancestry of Verebrates* is just a little rhyme pointing out the traces of our protochordate ancestry. In rather different style *Leptocephalus brevirostris* presents us in a wide sweep with the marvellous story of the breeding of our freshwater eels, after their migration across the Atlantic to the depths of the Sargasso Sea, and the two years drifting of the baby fry, carried back by the Gulf-stream and its branching water, to reach Europe's shores and rivers once more. He ends it with a fitting salute to his old Danish friends who made these now classic discoveries:

' Johannes Schmidt the leader, with ' Papa ' Petersen behind,

[1] Here he also refers to his views that these animals are paedomorphically derived from Doliolid larvae rather than from the larvae of sedentary ascidians.

Who made the 'Thor' and 'Dana' ships of fame for all mankind.

The Axolotl and the Ammocoete begins by describing very prettily the often uncertain neotenous condition of the Axolotl and the permanent neoteny of the perennibranchiate amphibia; and then, looking at the Ammocoete, he suggests that if *it* should become neotenous we should have to revise our ideas of the ancestry of the Lancelot (Amphioxus). This, as indicated above, p. 14, is the idea he arrived at independently of Stensio.

The last two poems are in a different category from the rest; they can hardly be classed as comic verse. In *An Oceanographer's Dream* he extols the great advantages of Bermuda as a site for a marine laboratory. Twice he had been there to work and just before the last war he was largely instrumental in obtaining financial support in this country for its marine station and the equipment and staffing of a research vessel. The final piece *To a Herring Gull* needs no explanation and has been published before in the *Oxford Magazine* (Feb. 6: 1920); permission to republish it here is acknowledged with gratitude.

Enough of introduction! I leave the stage for our zoological entertainer—the ever youthful Garstang the Jester!

THE AMPHIBLASTULA

and the Origin of Sponges

THE *Amphiblastula* is notable as a larval pioneer.
Who disregards all maxims of development each year:
Half flagellate, half amoeboid, he takes his silent plunge,
Adheres, and tucks his *front* end in, to build a baby Sponge.
Why does the *Amphiblastula* do things the wrong way round,
When hinder poles are recognised as vegetative ground?
The *Parenchymula* is worse, for he turns inside out,
His dermal cells being all inside, his gastral all without!

The answer's fairly simple: Sponges feed like Protozoa,
And haven't any enteron like other Metazoa;
Their ' gastral cells ' are clusters of primaeval Collared Monads,
And their bodies have no traces of muscles, nerves, or gonads.
Even their eggs are products of choanocyte divisions,
And therefore carry with them Choanoflagellate traditions:
Flagellate or amoeboid, with collars or without,
They still are Collared Monads—though certainly more stout.

In the early Ascon larva all the cells are first flagellate (The
Collared Monad's swimming form, as its creeping form is stellate),
Then the hinder cells progressively their flagella all draw in,
Becoming mere amoeboid balls predestined for the skin.
Fixation brings the collar-cells to a pile upon the floor
With the amoebocytes on top of them, extended more and more:
Then all together straighten out as a double-sided drum,
Flagella beat, and water-pressure bursts the osculum.

Were you a Collared Monrad, seated lightly on your stalk
With 3 or 4 around you, and should they start to talk
Of social reconstruction, with a fuller family life,
And greater home security, and less competing strife,
Could you do any better—from the Monad's point of view—
Than *Leucosolenia's* larva has managed here to do?
Had he regarded germ-layers and the doctrines taught in Jena,
Would that have made a better Sponge, or his procedure saner?

And as for *Parenchymula*, with his dermal cells inside,
Remember he is larger, with a longer course to ride!
To crowd the decks with cargo, when there's lots of room within,
By every navigator would be counted deadly sin.
In Maas's classic memoir one can see how finely drawn
Is the balance between cargo and locomotive ' brawn ':
For in Mobaxonid larvae, which with spicules load their bales,
A motor ring is added too, which evens out the scales.

A Sponge is no degenerate, then, from a loftier creation,
But a progressive Monad colony with differentiation.
The change is great, undoubtedly, as anyone can see,
But still there's continuity with the Collared Monad's tree.
The trees of Collared Monads are family erections:
The zygote and his progeny each build successive sections,
Until the slender candelabra cannot bear more lustres,
And the topmost generations sit unstalked in graceful clusters.

This forms the actual climax of the Collared Monad state:
Zygosis possibly ensues, and the Monads separate,—
Perhaps to start again their one colonial enterprise,
The branching common pedestal on which they all can rise.

The basal lines on which the Sponge from this estate has risen
Would clearly then appear to be a primary division
Between trophocytes to deal with food and sclerocytes to build,—
Which in the Amphiblastula we've seen to be fulfilled.

The ovum first pure collar-cells in front proliferates,
And then a generation of amoebocytes creates;
These build around the first a porous spicule-stiffened wall,
To which the collar-cells can cling instead of stalks for all.
No further dare we go until Hexactinellids yield
Their story of development, at present unrevealed:
And how they feed their ova, and if their chambers show
Detachment in their origin, as Monad clusters grow.

THE INVAGINATE GASTRULA AND THE PLANULA

A giddy little *Gastrula*, gyrating round and round,
Was thought to show the way we got our enteron profound:
A little whirlpool in its wake maintained a tasty store,
A pocket sank to lodge it all, and left a blastopore.

As a larval epigram this description earns a prize,
But as sketching adult ancestry can only win surprise,
And when you note all early orders fixed upon the rocks,
You feel a slight embarrassment, the first of many shocks.

The foremost inconsistency is the simple, solid fact
That the Hydrozoan larva, the mouthless *Planula*, is packed
With a jumbled mass of gastric cells that drop in, slow or fast,
And show no slightest cavity till the larval stage has passed.

The larva then becomes attached, and shortly stands erect,
Nourished by the yolky stores the inner cells eject:
Their shrinkage leaves a growing space, the early enteron,
Round which a layer of cells remains, and lines the outer one.

Some tentacles are sprouted then, say 2, then 4 and 8,
And not till all is ready does the mouth break through quite late.
If mouth and gut arose at first from one invagination,
What roundabout procedure's here! What needless complication!

Invagination surely is a thing of later date—
Procedure speeded up to suit the embryonic state:
The cells as loose irregulars build up the lower grades,
And yield but slowly, step by step, to organised brigades.

THE ORIGIN OF CNIDOBLASTS AND CNIDOZOA[1]

'TIS odd that Enterozoa should with Coelenterates begin,
With differentiated cells and a diplothelial skin,
For the Gastrula is clearly by a Blastula preceded,
And pelagic monothelial sires for this are sorely needed!
'Tis also true that Haeckel, when he looked around for one,
Could only '*Magosphaera*' find, which no one else has done!
He then appealed to *Volvox*, which the serious dearth reveals,
Since both are quite incapable of taking solid meals!

In short the germ-layer doctrine, which applies to later forms
Provides no guide to help us with the Protozoan swarms,
From which, unless we're all astray, the Metazoa rose,
And Sponges illustrate a stock which yearly from them grows.
As the collar-cell to Sponges, so the explosive cnidoblast
Is the keystone of Coelenterate evolution first and last:
Once admitted that their bodies have been built to give it scope,
Reconstruction should be easy—or is this a pious hope?

Between Sponges and Coelenterates strong barriers are set,
Which mark distinctive pathways from their origin—and yet
A difference at the outset does not common points expel,
And the building of a Polyp should run closely parallel.
But in Sponges food-collection and digestion are the same
As in the primal zygote, which we can even name:
While in Cnidozoa[1] these functions are completely separated,
And their agents in accordance widely differentiated.

[1] An alternative name for the Coelenterata (and one which Garstang preferred. A.C.H.).

Can any common sire be found for the simplest cnidoblast—
Which captures all the prey—and an epithelial myoplast?
For these in Cnidozoa are th'essential units of construction,
And without an answer there's no base for any sound induction.
A tentative reply could be: ' Euglenoid Flagellates! '
Their myonemes have found a rôle in both the later states,
And their ' gullet ', while it disappears from the epithelial cell,
By introversion might a simple cnidosac foretell.

Or are the ' Heterokaryotes ' to Coelenterates akin,
As retrogressive offspring from a common origin?
The cnidosacs, though nowadays arising introverted,
Like Acinetan suckers must at first have grown exserted.
The primal Acinetan may itself have been derived
From a colony of Euglenoids that had not quite contrived
To separate their cytoplasms when undergoing fission,
Thus setting up a unifying *Volvox*-like condition.

And before Euglenoids fell a prey to saprophytic thirst
They must have fed on particles, as Holozoa, first,
For otherwise some reason must be found that we can follow
Why Euglenoids have a gullet with which they never swallow.
In such a Holozoic stock of sessile Flagellates
The zooids round the edge would catch their food at greater rates,
But, being to their fellows joined by protoplasmic threads,
A normal consequence would be—differentiated heads![1]

Flagellates in the centre work their currents as before
And catch the finer particles, but the others aim at more;
With their gullets far extended they attack the larger fry,
Abolish their flagella, and suctorial methods ply.

[1] c.f. Hydractinia, Hydrocorallines, Disconanthæ.

With this achieved the Acinetans sought consolidation,
Their bodies and their nuclei underwent full integration:
But another line continued, multicellular in state,
And, with dimorphic units, was reserved for higher fate.

To argue how the diplothele condition was attained
At present would be premature, but a basis will be gained
If the prior question's recognised as not the double wall
But the origin of cnidoblasts.—First things first of all!

CONARIA AND CO.

CONARIA'S an *Actinula*, *Velella's* little daughter,
Reared by a Medusa nurse in dark abyssal water:
But Coney said: ' This dismal den is no abode for me!
I mean to rise, ev'n if I have to float a company! '

Her stiff and useless laval arms she thereon cast away,
And sprouted zooids round her, some 3 or 4 a day.
These fed her well, perhaps exceeding normal Carid doses
(Or was it an ' unsteady state ' in the rate of her osmosis?)

In any case, whate'er the cause, some gas got in her float,
And things began to happen which it scared them all to note:
Familiar phosphorescent glows went down or faded dimmer,
It seemed less cold, and now and then faint light above would glimmer;

Strange creatures sank, but Coney's stock was clearly on the rise
And soaring—through the Photic Zone—until with glad surprise
Day changed to night, and night to day, and with a shock the more
They broke the waving surface as a blue Siphonophore.

With an air-filled float and sail aslant upon a sparkling sea
Conaria merged her little stock in *Velella*, L.t.d.!

MÜLLERIA AND THE CTENOPHORE

JOHANNES MÜLLER's larva is the primal Trochophore
That shows how early worms grew up from fry in days of yore:
No drastic metamorphosis!—each youngster keeps her skin:
Her larval frills are not thrown off, but eaten from within.

A band of cilia on her frills is more than pretty trimming,
For, looping down beside her mouth, it feeds and keeps her swimming.
The simpler forms with central mouth bear one such flap alongside,
And one of these, *Pilidium*-like, waltzed by—but on the wrong side!

She bumped into a Ctenophore—a serious thing to do,
As sundry careless Copepods already sadly knew!
With scorn he said: ' You clumsy thing, why don't you learn to steer?
Had I not been preoccupied, this might have cost you dear.

Why do you spinning folk keep up this antiquated gait,
Which even flapper Polychaetes admit is out of date?
Their girdle is of course too straight for graceful evolution,
But you, with one more flexible, might follow my solution.'

Here Ctenie gave her such a show of costal navigation
That little plump *Mülleria* was lost in admiration:
Ahead, astern, above, below, he crossed and doubled round her,
Till she bewildered dropped her arms, as if about to founder.

' That's it! You've caught the trick! ' said Ctenie, ' Keep your arms well down!
Now lift the four loops up between, and raise them to your crown!
You've got a sort of sensory patch that, with a little care,
Might even grow an otolith and work them all from there! '

Mülleria faintly sighed and said: ' Indeed, you're very clever!
But you forget I haven't time for such prolonged endeavour!
The loops you want me now to raise are bound so very firm—
Besides I must be nearly due to turn into a worm! '

Then Ctenie said: ' To you, my dear, I wouldn't tell a fib! '
(And here he nudged what might have been her subsagittal rib)
' We never talk about it now, but once upon a time
My ancestors, like yours, were worms, and crawled about in slime.

' Ev'n now the tendency is there, though most of us are saner,—
And retrogrades we can't escape, as witness *Ctenoplana*!
But, having tasted joys above, our hopes could not be quelled:
We worked a revolution, and against our sires rebelled.

' Year in, year out, we looped our girdles higher up and higher,
Postponing thus the evil day of dropping to the mire,
Until at length the sun and shine achieved our great ambition:
While still aloft we laid our eggs, and broke the worm tradition!

'And now we're such a feature of these altitudes pelagic
That no one dreams our history has bordered on the tragic:
They even put a gloss upon our state hermaphroditic,
And class us with Coelenterates—delicious, if uncritic! '

But all these vistas conjured up proved too much for *Müllería*;
Her armlets shrank, she flattened out, showed symptoms of hysteria,
Her girdle split and disappeared, her eye-spots twinkled sad,
She gave a last, long, lingering look, and sank—a Polyclad!

1922 (or earlier).

THE ONCHOSPHERE

THE Onchosphere or Hexacanth was not designed for frolic,
His part may be described perhaps as coldly diabolic:
He's born amid some gruesome things, but this should count for virtue,
That steadily, 'gainst fearful odds, he plies his task—to hurt you!

He's very small, a mere pin's head, beset with six small hooklets,
Is whirled about by wind and rain through puddles, fields and brooklets;
But if a pig should swallow him, as many porkers do,
He's made a start with no mistake: he's on the road to you!

Again I say, don't blame the brat—he hasn't any head!
It isn't any fault of his—he wasn't painted red!
But once inside, he burrows through, and gropes his way about,
Then swells and sprouts a head at last, though this is inside out!

He's now a *Cysticerus* in the muscles of a pig,
With just a sporting chance of getting out to grow up big.
If you'll consent to eat your pork half-raw or underdone,
His troubles will be over, and a Tapeworm will have won:
He'll cast his anchors out, and on your best digested food
Will thrive, and bud an endless chain to raise a countless brood.

1922 (or earlier).

THE TROCHOPHORES

THE *Trochophores* are larval tops the Polychaetes set spinning
With just a ciliated ring—at least in the beginning—
They feed, and feel an urgent need to grow more like their mothers,
So sprout some segments on behind, first one, and then the others.

And since more weight demands more power, each segment has to bring
Its contribution in an extra locomotive ring:
With these the larva swims with ease, and, adding segments more,
Becomes a *Polytrochula* instead of *Trochophore*.
Then setose bundles sprout and grow, and the sequel can't be hid:
The larva fails to pull its weight, and sinks—an Annelid.

In this way fare *Phyllodoce*, but *Nereis* can beat her:
She gives each egg some extra yolk to hatch as *Nectochaeta*.
The simple stage with prototroch is by-passed in the eggs,
And each when hatched has three good pairs of parapodial legs.
With these it paddles on in jerks, and could it change its skin
Would almost be a *Nauplius*, its very near of kin.

But to extend their life afloat as long as e'er they can,
Young *Spio* and *Mitraria* pursue another plan:
In front of all the segments which of setae still are bare
On either side they sprout a tuft of shining bristly hair,
And these erected suddenly form a kind of double shield,
Which must some vital purpose serve, in our next to be revealed.

MITRARIA'S FAN DANCE

AS when on quiet summer eves
Where sunbeams filter through the leaves
The Mayflies join their revels,
Mounting where sunshine most prevails
And then on balanced wings and tails
Floating to lower levels,

So may one see these larval mites
Press upwards towards the brighter lights
And, for the counterpart,
Flick out, like fans, their tufts extended
And, as on parachutes suspended,
Fall slowly to the start.

Is it, as there, just playtime zest?
More probably, be it confessed,
Their mode of food collection:
For diatoms crowd up aloft,
And movements that are smooth and soft
Can best escape detection.

Now equatorial waters warm
Present against a sinking form
A much reduced resistance:
A North Sea creature falling there,
Without some compensating care,
Would drop through twice the distance.

So tropic *Mitraria* drifting down
There spread not bristles but a crown
Of broad and gleaming blades,
Which some say give protection martial,
But we—to seem not over-partial—
Will call fanfaronades!

THE BALLAD OF THE VELIGER
OR
HOW THE GASTROPOD GOT ITS TWIST

THE Veliger's a lively tar, the liveliest afloat,
A whirling wheel on either side propels his little boat;
But when the danger signal warns his bustling submarine,
He stops the engine, shuts the port, and drops below unseen.

He's witnessed several changes in pelagic motor-craft;
The first he sailed was just a tub, with a tiny cabin aft.
An Archi-mollusk fashioned it, according to his kind,
He'd always stowed his gills and things in a mantle-sac behind.

Young Archi-mollusks went to sea with nothing but a velum—
A sort of autocyling hoop, instead of pram—to wheel 'em;
And, spinning round, they one by one acquired parental features,
A shell above, a foot below—the queerest little creatures.

But when by chance they brushed against their neighbours in the
briny,
Coelenterates with stinging threads and Arthropods so spiny,
By one weak spot betrayed, alas, they fell an easy prey—
Their soft preoral lobes in front could not be tucked away!

Their feet, you see, amidships, next the cuddy-hole abaft,
Drew in at once, and left their heads exposed to every shaft.
So Archi-mollusks dwindled, and the race was sinking fast,
When by the merest accident salvation came at last.

A fleet of fry turned out one day, eventful in the sequel,
Whose left and right retractors on the two sides were unequal:
Their starboard halliards fixed astern alone supplied the head,
While those set aport were spread abeam and served the back
instead.

Predaceous foes, still drifting by in numbers unabated,
Were baffled now by tactics which their dining plans
frustrated.
Their prey upon alarm collapsed, but promptly turned about,
With the tender morsal safe within and the horny foot
without!

This manoeuvre (*vide* Lamarck) speeded up with repetition,
Until the parts affected gained a rhythmical condition,
And torsion, needing now no more a stimulating stab,
Will take its predetermined course in a watchglass in the lab.

In this way, then, the Veliger, triumphantly askew,
Acquired his cabin for'ard, holding all his sailing crew—
A Trochosphere in armour cased, with a foot to work the
hatch,
And double screws to drive ahead with smartness and
despatch.

But when the first new Veligers came home again to shore,
And settled down as Gastropods with mantle-sac afore,
The Archi-mollusk sought a cleft his shame and grief to hide,
Crunched horribly his horny teeth, gave up the ghost, and
died.

1928.

ECHINOSPIRA'S DOUBLE SHELL

A HANDSOME 6-lobed Veliger, to wit *Echinospira*,
Constructs his house with double shell to baffle the enquirer,
An outer like an Ammonite's, symmetrical, capacious,
An inner fitted to his back, flattened and less spacious.

While the inner shell is smoothly curved, the outer's decorated,
Like an Ammonite's it's ' involute ', its aperture ' rostrated; '
It has spiral rows of ' tubercles ', or ridges in default,
And its ' venter ' bears a ' channel ', like *Hoplites* of the Gault.

Echinospira sets this riddle to the students of Conchology
To make them pay attention to the doctrines of Morphology:
And this is how he poses it: ' The Ammonite's old shell
From time to time was portioned off, to make it fit him well.

' The inner shell around his hump was " visceral ", like mine;
His outer shell also agreed: it's " pallial " in fine.
We differ just in this: his inner shell was fixed by suture,
While mine is truly portable, and useful for the future!

' So let us sing in fitting terms an *entente cordiale*,
Observing in its proper place the *torsion viscerale*:
My outer shell's a " relic " of my Ammonitic traits,
My inner is a tribute to my clever modern ways! '

. . . .

Echinospira's argument has a touch of the satiric:
He tells the truth, but not the whole, to suit his playful lyric.
If you could see, as I have watched his growth and transformation,
Behind his riddle you would read amazing adaptation.

Lamellaria is a Prosobranch who spreads for prying eyes
A deceptive over-mantle in the motley coloured guise
Of various Synascidians, on which she likes to feed,
And in crypts of which she lays her eggs, when the time has come to breed.

Her hidden shell, though thin and flat, supports the tender skin
That roofs her mantle-chamber with her gill and things within:
Her shell has doubtless been preserved for this one rôle it plays,
While completely lost in Nudibranchs, so like in other ways.

Confirmed by Pelseneer, M. Gaird long ago
Observed two shells successively formed by the embryo:
The first arose correctly, and the mantle-edge embraced it,
The second wedged itself beneath, on the visceral hump which faced it.

During larval life the two shells grow at very different rates,
And, divergent in their sources, are contrasted in their fates;
The inner, purely visceral, grows slowly with the body,
The outer, purely pallial, is precocious, smart, and shoddy.

But most significant of all, the quick-growing outer shell
Demands the mantle's increase at an equal rate as well:
To swim and feed it must to the outdrawn aperture extend,
Along with the great velum on which their lives depend.

'Tis then, with edge reflected o'er the apertural brim,
It makes its shell additions, all parallel and trim:
And where the aperture expands, in the middle of both sides,
A lateral ridge runs backward, and the coil outside divides.

The ridge serves as a building line to which successive whorls
Attach themselves as they extend their logarithmic curls,
Until some centrifugal force upsets the normal scales,
And a vestibule runs off the track, outside the larval rails.

This change of aperture ushers in the larva's final phase
(A ' gerontic reminiscence ' of old Ammonitic days!),
And the last thing that the larva does ere the adult takes the floor
Is to throw away an operculum that never closed the door.

If useless, then why did he keep it all the previous time?
The larva's eccentricity here borders the sublime!
From the way it's formed, his house should be a coiled and tapering tunnel,
But when you take a peep inside, there's nothing past the funnel!

In accordance with its building line, it embraces in its curl
The roof and half the sides of the previous body-whorl:
But in this *Echinospira* saw a waste of living space
And determined, as new coils rolled in, their roofing to efface.

He set his sharp operculum as a carpenter his plane,
With orders, as he glided out, regardless of the grain,
To slice away all obstacles to a horizontal level,
Radially outwards to the apertural bevel.

And every time the mantle-edge from the aperture withdrew,
It left a trail of mucilage along the passage new,
Which, hardening into strings and sheets sufficient for his guidance,
Smoothed all the jagged ends of the former floor's subsidence.

At Plymouth once I realised what had been a cherished wish,
When, with various *Echinospiras* swimming gaily in a dish,
Was one who lay upon his side in seeming tribulation,
And the microscope revealed him in the throes of transformation.

His mantle was retracted and writhing as in pain
With convulsive movements everywhere but no apparent gain,
Until a gradual shifting of the struggling mass revealed
That the whole was being reversed upon the inner adult shield.

In 5 minutes this was evident, in an hour it was complete,
Except above the shield a rounded hole, soon small and neat.

Next morning scarce a trace was left, and with a Trovatore aria
I welcomed in my capsule a perfect *Lamellaria*!
In the darkness he'd crept out of *Echinospira's* vestibule,
Depositing as a souvenir his larval carving tool!

. . . . .

Now let us give this complicated tale a brief review,
For to the older stories we can add an aspect new:
The two shells are not ' false ' and ' true ', the false one first to
start,
But both are true and false alike,—one true shell split apart.

The glandular secretive disk which lays the shell's foundation
Should normally add successive layers to its earliest formation:
But the central part in this case failed to keep in contact close,
And its next secretion merely formed another primary dose.

Two mutilated rudiments thus followed this mutation,
They should have clung together and shared a common
destination,
With visceral layers underneath and pallial around,
By the mantle-margin's active edge harmoniously bound.

But by the breach the first half lost all visceral contributions,
And the second got no benefit from the mantle-edge effusions:
So each grew up as best it could, independent of the other,—
Pallial and visceral counterparts unbound to one another.

If we call the younger ' Jacob ', and the older one name ' Esau ',
Between the two the mantle plays its curious game of see-saw:
As the elder, Essau gets it first, and enjoys a gay sea-spin,
Then Jacob steals the whole estate, with the spacious cloak thrown in!

In a precious sheaf of memoirs Marie Lebour has shown
That in this respect the Lamellarians do not stand alone:
The embryoes of Cowries in particular should tell
The earliest of all stages of the duplicated shell.

The larval shells of *Trivia* and *Erato*, helicoidal,
Strikingly contrast with *Echinospira's* ammonoidal:
This means a greater influence of the body's visceral twist,
Which *Echinospira's* shell by its swifter growth has missed.

Thus *Echinospira's* symmetry is an entirely new creation,
With Nautiloids and Ammonoids having no relation:
And *Lamellaria's* ancestry to the Cowries can be traced
In spite of common characters having largely been effaced.

But, beside the doubled shell, another heirloom they've bequeathed
In the freely flowing mantle which their stony shells ensheathed:
And, whatever its first function, it was clearly not unwise
To have a means of camouflage, when needing a disguise.

Lamellaria, inheriting, was not content with this,
And that is how she came to change her metamorphosis,
Which replaced the coiled ancestral shell by a flatter, simpler shield,
At the outset and for always by a mantle quite concealed.

15/11/43.

THE NAUPLIUS AND PROTASPIS

THE *Nauplius* is a wobbly thing, a head without a body:
He flops about with foolish jerks, a regular Tom-noddy.
Some said he was an ancestor, but others said: ' What, HIM?
He's just a *Nectochaeta* with Crustacean skin and limb! '

But *Nectochaeta's* segments all are clearly demarcated,
While in *Nauplius* all five segments with the head are integrated,—
So solidly that boundaries above can not be traced,
And maxillary appendages for some time are effaced.

The simplest *Nauplius* thus has left simplicity behind,
And to the gap forever we should have to be resigned
Had not the ancient Trilobites—Crustacean ancestors—
Behind them left their larval shells on Palaeozoic shores.

These little disks, ¾ of a millimetre wide,
An almost perfect missing link between the two provide:
A quinquepartite axial tract defines the compound head,
And flange-like cheeks of varied form on either side are spread.

Of actual appendages no traces have been found—
Nor could more be expected than streaks upon the ground—
But the transverse equidistant lines that cross the axial tract
Are of uniform appendages as eloquent as fact.

The *Protaspis*, like the *Nauplius*, then a groove behind displays,
Which separates a telson from the head, and so, by ways
Familiar in Annelids, new segments, one by one,
Are interpolated till the whole development is done.

KENTROGON

KENTROGON'S a crippled imp, uncanny, lacking kin,
The 'stabbing seed' a *Cypris* made by shrinking from her skin,—
A *Cypris*? Nay, a fiend that borrowed Cyprid feet and mask,
To cast them off when he had plied his victim-hunting task.

From the cover of a *Nauplius*, one of *Sacculina's* daughters,
Is launched the Cyprid coffin-ship to break beneath the waters:
Then *Kentrogon* emerges from the wreckage on the floor
And throws his crafty life-line to carry him ashore.

Through the antenna of the *Cypris* on the hairlet of the Crab
His fine proboscis travels and inflicts the needful stab:
Then, gathering all the salvage that he's rescued from the rout,
He slips along the tunnel of his own projecting snout.

So *Kentrogon*, like Charon, carries souls from light to dark,
Himself at once both ferryman and passenger and bark,—
A Phoenix all perverse who on his desperate day of doom
Refanned his flame to start afresh within a living tomb.

1922 (or earlier).

ISOPOD PHYLOGENY

SING a song of six legs, a new phyletic stage!
Four and twenty Isopods cradled in a cage:
When the cage was opened, out they ran to play—
Wasn't it a jolly thing to have a jolly day!

Mother rocked the cradle between her stegopods:
The youngsters ran about her seven pereiopods:
When they found she'd one pair more than they themselves,
They called a hasty conference on oöstegal shelves.

MacBride[1] was in his garden settling pedigrees,
There came a baby Woodlouse and climbed upon his knees,
And said: ' Sir, if our six legs have such an ancient air,
Shall we be less ancestral when we've grown our mother's pair? '

1922 (or earlier).

[1] cf. *Textbook of Embryology*, 1914: ' They (i.e. the larvae of ***Portunion***) resemble the young of normal Isopods when they leave the brood-pouch, and not even the most determined opponent of the recapitulation theory could deny their ancestral significance,' p. 219.

THE MILLIPEDE'S EGG-TOOTH

THE hatching of a Millipede brings curious things to light:
The embryo within its shell is curled up snug and tight
Enclosed inside an inner skin with a thorn upon its neck,
Whose task it is to pierce the shell, as chicks their prisons peck.
What is this extra covering that thus comes into view?
An heirloom from antiquity here blended with the new?
Another ' Nauplius-coat ' around another embryo,
The same that Peracarids on their cradled babes bestow?

'Nonsense!' I hear it lightly said. Yet the facts are much the same:
An Arthropod with yolky eggs, whatever be its name,
Must own aquatic ancestors that once laid smaller eggs,
And hatched as tiny larval forms with many fewer legs.
So to those fry prophetic of our Millipedes and Spiders
(And Centipedes and Insects surely cannot be outsiders)
Above their necks, behind their heads, we'll venture to assign,
As their outstanding feature, a median dorsal spine.

Now let us turn to Trilobites, those monarchs of the prime,
Which left their larval shells by scores in Palaeozoic slime,
And, passing o'er the later forms with aspect smooth and formal,
In Cambrian shales we find our types in armour quite abnormal.
The Mesonacids bristle with segmental dorsal spines,
And scores of highly varied forms mark out a dozen lines:

And ev'n when body-spines have retrogressed beyond dispute
A prae-thoracic spine stands out the critics to confute!
Witness *Callavia* broad and plain, *Pagetia* specialised,—
Two Cambrian forms as far apart as any then devised:
Yet each with this cephalic spine is strikingly bedecked,
With vestiges of others not so easy to detect.

For adult Trilobite defence these spines were never meant.
There were no crabs or fishes then to prowl on mischief bent!
The problem was a larval one, to float for distribution
When seas were warm and salts were yet in very weak solution.

So Mesonacid larvae were built intensely spiny,
Thus offering larger surface for contact with the briny:
The spine upon the head came first, and, disappearing last,
Gave Proto-tracheates a tool when swimming needs had passed.

THE TRILOBITES AND AFTER

ERE Snowdon raised its lofty head above the Cambrian Sea
King Paradoxus called his sons—and many sons had he.
He looked around—and looking, was an action quite profound—
For his eyes were very rigid: 'twas his body that turned round.

'My sons,' he said, 'our neighbours, whether Brachiopod or
Cystid,
Contentedly upon their stalks on sea-soup have existed,
The Graptolites that drift around have never done us hurt,
And we, detritus-feeders, have survived the dust and dirt.

'But a uniform environment of blue-black marl and mud
Has not induced variety, and variety is good.
The sea is getting deeper and the land is rising higher,
We're doomed unless we take some steps to get above the mire.

'I know we're not for enterprise ideally designed:
We're over-squat, with feeble legs exactly of one kind:
But if some of you would go ashore and change your respiration,
You might improve on what you see, and work a reformation.'

The Paradoxus joined his sires, and the shore went up and
down,
And every year adventurous sons would frizzle up or drown,
Until they learnt the tidal ways and from the sun took shelter,
Instead of getting all bemused and wandering helter-skelter.

And after several million years and countless oscillations
Beside a geyser lake there met two parted generations:
Lepidocaris in the water was paddling near the shore,
And *Arthropleura* on the bank peered over more and more.

' Hallo!' said Carrie, ' surely you are one of my lost cousins? '
' O.K. ' said Arty, ' in the sea I knew your lot by dozens.
But how you're changed! Have you been ill? Unless the light's
too dim,
Your pleura seem to droop a lot, and you seem very slim.'

Carrie thereon indignantly dashed off across the lake,
And Arty realised he'd made a very big mistake.
But Carrie on returning said: ' You needn't talk of slimming!
What I have done, as now you see, has been for better swimming.'

' Sorry, my dear,' then Arty said, ' I made a stupid blunder.
Old Paradox would hail you as a Trilobital wonder!
The way you've made your exopods work freely with your endo's
Is marvellous, a triumph of evolutional crescendos.'

' Indeed it wasn't easy,' Carrie then to him admitted,
' Our exopods, 'twixt endopods and pleura tightly fitted,
Had barely room to oscillate within their latticed cage,
And the endopods's slow movements put me daily in a rage!

Then once, when trying to spurt a bit with those horizontal oars,
They tangled with the exopods above the pleural floors;
So recklessly I wrenched them out and worked the lot as one,
And really thought I'd done the trick—and nearly was undone.'

' Why, what do you mean? ' asked Arty. ' You got them free at
last,
And I'm sure no other Trilobite e'er travelled half as fast.'
' Oh, now's all right,' said Carrie, ' but at first 'twas touch and go!
My endo's wouldn't feather, and my exo's wouldn't row!

' So awful trouble followed with my pleural respiration,
I had to work my antennae hard to avoid asphyxiation.'
' I'm jolly glad you managed,' Arty said, ' I wish that I
Had never left the water, for I might have helped near by.'

And then he told her how he'd lived for months upon the shore,
Exploring every crevice that held water on the floor,
Until by coming out by night and keeping damp with spray
He almost was acclimatised to breathing any way:

'But what's the good? Compared with you, I'm rubbish on the ground,
Condemned to keep below by day and never move around.
What have I gained by leaving the old mud-banks after all,
When now I'm doomed for ever just to burrow and to crawl?'

But Carrie, as Arty sadly spoke, grew more and more excited,
And, when he'd finished, said: 'My lad, you're far from being blighted;
You left the mud-banks after me, but have really done much more,
For you've won a new environment. Your triumphs lie before!

'Like me, I see, you've sprung a hinge upon your pleural flaps!
With me they're bound to dwindle, but with you (who knows?) perhaps
When your method of air-breathing has made a further slight advance
You'll lengthen out those pleural wings and gaily with them dance
In wider realms of airy space than I shall ever know!
Go, therefore, Arty, cheerfully—with my best wishes, go!'

She swam away, and when she'd gone, with a fearful rumbling roar
The boiling geyser burst again and flooded lake and shore—
Wee *Lepidocaris* was lost, then and for evermore,—
  At least till Mr. Scourfield found and pieced her sad remains,
And told the tale we here re-tell in slightly altered strains.

ACTINOTROCHA

THE larva of *Phoronis*, known as *Actinotrocha*,
Plays better than most a good hand at poker.
Bluffing you first as a new *Bipinnaria*
(With relations behind to her cousin *Tornaria*),
She shows you praeoral and postoral bands,
Divided above into separate strands;
And when in your face you betray a few qualms
(The former being simple, the latter with arms!)
And your eyes glance again at her praeoral band,
She turns inside out, and discloses her hand!
Such behaviour by larvae is quite *comme il faut*,
And you say, when she's finished, 'I'd much like to know
What she really means when she tries to convince us
Bipinnaria's a young *Pro-cephalodiscus*!'

CYPHONAUTES

THE *Cyphonautes* larva wears a cocked hat for a shell,
With involuted sucker and a vestibule as well:
His mask, a very thin disguise, may be a playful joker's,
But whose the visage underneath if not *Actinotrocha's*?

ECHINODERM LARVAE

AND THE ORIGIN OF QUINQUERADIAL SYMMETRY

BIPINNARIA should a tale unfold when she gives up life afloat
And anchors by her bows, Crinoid kinship to denote:
She might by just a twist reveal how Starfish went astray,
But no! She merely buds one out, and fades herself away.

This is highly tantalising, but at least *Auricularia*
(Her sister who in course of time becomes a *Cucumaria*)
Is kinder and throws welcome light on some important things,
When breaking up her feeding band for locomotive rings.
Admittedly odd families some episodes must hide,
But she displays possessions that feed a proper pride,
And in her circumoral band shows upper class connections,
Which her sister, just for anchorage, dismembers into sections.
Their two twin sisters all conventions openly defy,
And drift about their business as simple *Plutei*:
Contentedly upon their backs, oblivious of scoffs,
They stretch their circumoral pits into spacious feeding troughs.

But, in spite of all their differences in form and modes of swimming,
A constant pattern underlies their circumoral trimming:
The pit itself expands each side like a transverse figure 8—
Two tiny pools connected by a narrow oral strait.
And each is fringed by brachioles approximately 5—
A number too suggestive for mere chances to contrive—
Their feeding streams converge upon the circumoral floor.
And all together constitute a bilobed lophophore.

Since the larval lophophore of *Actinotrocha* is supplied
With a pair of ' collar ' coelomes, these can scarcely be denied
To former Echinozoic fry, which hydraulic pressure needed
To support their lateral armlets ere a skeleton succeeded.

At the ensuing metamorphosis all larval frills are lost,
The eventual adult body's built at the larval body's cost:
Quaintly reminiscent of the birth of Adam's bride,
The Starfish as a bud grows out from its larva's own left side.
To this strange rule a fortunate exception still survives:
Auricularia a transitional pupal state contrives,
In which the mouth and pharynx undergo a double twist,
First sideways to the left, then to the apex of the cyst.
The twist involves no torsion of the actual body wall,
But a simple shifting of the mouth alone and once for all.
If we reconstruct the lophophore, the fact is then revealed:
The mouth has turned aside within the circumoral field.

The right half, losing value, degenerates with the torsion:
Its hydrocoele accordingly undergoes complete abortion.
And, whereas in the larva the bilobed lophophore was ventral,
The left lobe, now around the mouth, becomes terminal and central.
'Tis known from countless fossils that the quinqueradial change
Was first achieved within the circumoral feeding range:
But how or why has not been clear, so it seemed worth while to try
If the explanation might be found in the structure of the fry.

THE PENTACRINULE

I FOUND a pretty Pentacrinule
Still without the arm and pinnule
That adorn an *Antedon:*
Desiring much to see her face,
I said, ' Coy creature, yield with grace,
And this slight favour grant a don! '

She split, and to my ardent gaze
Displayed her mouth and radial ways
With tentacles beside them:
But as I viewed her dainty fringes,
Five trap-doors on five broad hinges
Promptly dropped to hide them.

Annoyed, I said: ' You naughty minx,
They told me you preserved the links
With early Crinoid nature.
Your oral plates you've undermined!
I'm sure they weren't as flaps designed.
You've changed a primal feature!

'And other things I find amiss:
A calyx with no radials, this . . . !
Oh, Recapitulation!
A finless tadpole, I confess,
Caricatures ancestry less
Than your sad aberration.

' But I'll not blame you, *Pentacrinus*:
Haeckel's catchwords still entwine us,
Parrot-like we talk!
And, though a vestibule has cleft you,
One old vestige still is left you
Till you quit your stalk! ' 1922 (or earlier).

TORNARIA'S WATER-WORKS

OF ALL the sea-jellies that dazzle the eyes,
Whose motions delight and whose forms surprise,
Two are outstanding, the gay Ctenophore
And the modest *Tornaria*: these hold the floor.
For symmetry, fitness, and manifold graces
Their eminence shines in the watery spaces:
Whether waltzing or gliding or poised quite still,
No others display such perfect skill.
But if I must one for the other relinquish,
And the premier star of the pair distinguish,
From a standpoint of sheer architectural beauty
A naturalist must, I think, feel it a duty
Of all the sea-jellies that dazzle the eyes
To give to *Tornaria* the highest prize.

Though built of mere jelly as broad as it's high,
But solid and firm as the lens of an eye,
It carries the curves of a stately bell
And its movements in stateliness none can excel.
At the top of its head are tiny red eyes
By which it can tell when to sink or to rise,
Which it does by a terminal flagellate ring
That circles it round in a clockwise spin.
It changes at will its specific gravity.
By adjusting the water in its hydrocoele cavity,
Which is fixed by a string to its apical pole
And opens outside by a left-sided hole.

Its mouth in the middle of the ventral wall,
O'erhung by a canopy broad and tall,
Is fed by two gutters that merge from its flanks,
Which are hollowed to serve as collecting tanks.

Each of these on approaching the apical pole
Forks like an anchor—but hollow, not whole!
And the anchor-shaped hollows on either hand
Are fringed by the circumoral band.
And bounding the furrows wherever they turn
Is a broad raised field in front and astern,
That fills up the gaps with elegant coils,
Embossed like a pair of carved trefoils.

The band seems at first to be subdivided,
But in fact it is not, though its loops have collided:
They fuse with each other round the apical plate
In which the two eyes are imbedded sedate.
To strengthen the force of the downward beat
Where ciliate band and furrows meet,
The band is produced into fine undulations,
Which multiply tenfold the small vibrations.
Thus the surface is everywhere furrowed and ridged
Like the crust of the earth, though much abridged!
And over it streamlets are made to flow—
Though all under water—from high to low:
But once in the valleys, the streams turn south,
And so to the gutters that lead to the mouth.

As windmills the water drain from the land,
And no notice taken, the ciliate band
Is the silent power of these marvellous works
In which much of *Tornaria's* beauty lurks.

OIKOPLEURA, JELLY-BUILDER

OIKOPLEURA, masquerading as a larval Ascidian,
Spins a jelly-bubble-house about his meridian:
His tail, doubled under, creates a good draught,
That drives water forward and sucks it in aft.

A filter in front collects all the fine particles—
Micro-flagellates and similar articles—
Which pour in a stream through a jelly-built tunnel
Into his mouth and its mucilage funnel.
The funnel begins with his endostyle gland,
Which flicks mucus up to his circular band:
The stream through his mouth trails it out into threads,
And the whole is rotated as fast as it spreads.
In effect this rotator's a neat centrifuge
That lets out the water and keeps in the ooze:
The water's sucked outwards by paired water-wheels,
The residue serves him with plentiful meals.

Now although *Oikopleura* sits by himself
In the midst of his house on a jelly-built shelf,
He's firmly attached in front by his snout,
And never lets go till his house wears out.
But his body behind is completely free
And bathed by the water that comes from the sea
Through two lattice-windows let into the walls,
Which limit the size of incoming hauls.
Into this water-space the effluents flow
That start from the spiracles' outward throw:
And lest water-pressure the bubble should burst,
A tubular valve in front blows first.

What shall we say of this marvellous creature
Who breaks all the rules by his composite nature?
The puzzle increases the more it's observed
How far from the track of his fellows he's swerved.
When his jelly-house starts as a lump on his back,
His tail is the finger that stretches it slack:
He probes with its tip between body and test
And loosens the parts which too closely are pressed.
Then, after windows and traps are all ready,
The tail pops inside and, with motions more steady,
Sets the pump working, the water streams in,
The jelly-house swells, and the fishings begin.

We believe we can satisfy any scrutator
That anatomy, house, and pharyngeal rotator
Are pure Doliolid in all their relations,
With highly original specialisations.
His tail is the problem and also the base,
For nothing will work if this you erase:
It seems that, from lack of metamorphosis,
He's larva and adult in half and half doses.

THE ANCESTRY OF VERTEBRATES

GILL-SLITS, Tongue-bars, Synapticulae,
Endostyle and Notochord: all these you will agree
Mark the Protochordate from the fishes in the sea,
And tell alike for them and us our lowly pedigree.

Thyroid,[1] Thymus,[2] Subnotochordal Rod[3]:
These we share with Lampreys, the Dogfish and the Cod,—
Relics of the food-trap that served our early meals,
And of Tongue-bars that multiplied the primal water-wheels.[4]

[1] Endostyle [2] Tongue-bars [3] Epipharyngeal Groove.
[4] Ciliated Stigmata.

LEPTOCEPHALUS BREVIROSTRIS

THE LARVA OF THE EUROPEAN EEL, *Anguilla Vulgaris*

LEPTOCEPHALUS of the Gulf Stream, the larva of the Eel,
Like a willow-leaf of clearest glass, set edgewise for a keel,
With a pair of eyes astride the stalk and a tiny cleft for jaws,
That wanders for 3,000 miles, two years without a pause.

We've known that other fish to distant spawning grounds would go,
Whence advantageous streams for drifting fry would tend to flow:
But these would serve mere fragments of a continental coast,
And gather fish from ranges of few hundred miles at most.

Yet from Iceland to Morocco and the Mediterranean lands
The silvered eels from every stream form vast autumnal bands
That in the sea converge to one far distant single goal,
Halfway between Bermuda and the Leeward Islands shoal.

Here the Atlantic deepens to its uttermost abyss
(Olympus piled on Etna here would still the surface miss!)
And over it th' assembled eels their bouyant eggs set free,
And never re-appear: they've fulfilled their destiny

Warm currents from the Canaries and the Equatorial Belt
Sweep broadly on above this spot and with the Gulf Stream melt,
And, swirling round Bermuda, set the whole Sargasso Sea
Gyrating like a whirlpool of gigantic majesty.

Blocked in the west, the rising waters pour their surplus east
To northern lands which hail its warmth as a recurrent feast:
It fills the Bay of Biscay too, and the thirsty Midland Sea,
Thus bathing all the coastal lands from which the eels went free.

Their fry, mere glassy filaments 6 mm. long,
At first the quiet underzones of the surface waters throng,
Like willow-leaves expanding then, their muscles spread each side,
And, beginning now to undulate, they rise into the tide.

Here they're wafted to and fro, and eddies circulate them,
The fringes of the current system widely separate them:
But, hatched in Spring, the bulk attain an inch in length by June,
And, one year on, to twice that length and quadrupled depth have
grown.

As active now as Whitebait, and swimming on in schools,
To get the best of current they have doubtless learn't the rules.
Of nosing from the weaker sides to where the major force is,
Thereby securing lodgement in the stronger water courses.

So, in spite of all the eddies at the outset to delay them,
In time they strike th' Atlantic Drift where nothing more can stay
them:
By Midsummer as 2-year-olds they've almost reached full size,
3 inches long, 3 quarters high, and crowned their enterprise.
They glide in now by millions o'er the Continental Shelf:
A pause ensues, and each begins to feel ' not quite himself '.
He's lost his hearty appetite: for months he shrinks a deal,
And finds himself, when 3 years old, an Elver or Glass-eel!

'Tis sad but true, he loses now all maritime ambitions.
His only aim's to settle down to all the old traditions;
To find some nice salubrious ditch or muddy-bottom'd stream,
Where he may fatten quietly and of his youth's adventures
dream!

. . . .

All honour to the Danes who solved this age-old mystery,
Who, step by step, and year by year, revealed the history:
Johannes Schmidt the leader, with ' Papa ' Petersen behind,
Who made the ' Thor ' and ' Dana ' ships of fame for all mankind.

THE AXOLOTL AND THE AMMOCOETE

AMBLYSTOMA'S a giant newt who rears in swampy waters,
As other newts are wont to do, a lot of fishy daughters:
These *Axolotls*, having gills, pursue a life aquatic,
But, when they should transform to newts, are naughty and erratic.

They change upon compulsion, if the water grows too foul,
For then they have to use their lungs, and go ashore to prowl:
But when a lake's attractive, nicely aired, and full of food,
They cling to youth perpetual, and rear a tadpole brood.

And newts Perennibranchiate have gone from bad to worse:
They think aquatic life is bliss, terrestrial a curse.
They do not even contemplate a change to suit the weather,
But live as tadpoles, breed as tadpoles, tadpoles altogether!

Now look at *Ammocoetes* there, reclining in the mud,
Preparing thyroid-extract to secure his tiny food:
If just a touch of sunshine more should make his gonads grow,
The Lancelet's claims to ancestry would get a nasty blow!

1922 (or earlier).

AN OCEANOGRAPHER'S DREAM

(with apologies to Tennyson)

I BUILT my soul a stately treasure house
  Beside the rolling ocean's swell,
With cunning apparatus to disclose
  The gems below that dwell,—

Those living gems ' of purest ray serene,'
  Oft frail, fantastic, strange and rare,
Whose lives sheer mysteries have ever been,
  But need not be—with care.

'Twas not in marbled luxury it grew,
  With spacious courts and colonnades;
A boat and cooling tanks were most in view,
  And tackle of all grades.

An isolated pinnacle I chose,
  Five hundred miles from every shore:
Sheer from th' Atlantic's deep abyss it rose,
  Three miles in depth or more.

Around it limpid dark blue water flows
  From distant equatorial seas:
It feels, above the pulse of melting snows,
  The Gulf Stream's vagrancies.

A sturdy, broad-beamed, fishing boat I planned,
  Equipped with motor winch and screw,
For quick excursions out and back to land—
  A man and boy for crew.

The zone of waters which concerns us most,
  Where mighty currents change their beat,
Can here be sampled three miles off the coast
  To full 3,000 feet.

BERMUDA, watch-tower of the Sargasso Sea!
  Its coral cap twelve miles across,
On this one spot my treasure-house would be,—
  Its absence the world's loss.

4/10/43.

TO A HERRING GULL

ON TIME'S CHANGES[1]

BOLD Sea-mew—you whose soaring flight
Inspires my envious Muse—
Pray, with this compliment polite
My liberty excuse

In asking what bird-politics
Urged you the seas to roam,
Whose spotted eggs and speckled chicks
Proclaim a moorland home.

Have you those upland joys forgot
By oozy bogs and rills—
Your Plover-cousins' happy lot
Among the fern-clad hills?

When Curlews flute their rippling call
And hasten to the moors,
Does memory then exert no thrall?
The voice of Spring no lures?

Was it perhaps that, as with men,
Succession by entail
Left Blackcap all the ancestral fen
And you—to face the gale?

You're on the brink! Remember, if
The land you quite forsake,
Your sailor brother on the cliff
Is now a Kittiwake,

[1] Reprinted from the *Oxford Magazine*, February 6, 1920.

And never, ev'n in winter gales,
  May to the fields repair,
To eke out there with grubs and snails
  His interrupted fare!

And all young Kittiwakes are vowed
  To go at once to sea:
No playing on the sands allowed,
  Despite their pedigree!

Nor may they don the brown-flecked dress
  Which your chicks wear to-day:
Young Kittiwakes must acquiesce
  In uniforms of grey!

Alas, what boots it? Fish and shrimps
  Monopolize your thought;
Nor dress, nor play, nor Spring-time glimpse
  Of ancient home counts aught!

You'd be a sea-bird? Well, I know,
  Persistence wins at last;
And, though your progress may seem slow,
  You're really moving fast!

True fishing-birds of course deride
  Mere puddling on the shore
And snapping flotsam from the tide—
  For deeper goes their lore!

Yon old black Cormorant on the rock
  Scoffs at your levity:
Your high top-hamper gives a shock
  To one who knows the sea.

Hull-down he swims: you merely float,
  A cork upon the waves;
Astern no storm-tried fishing boat
  Rigs out a pile of staves!

And then your clamour, your wild screams,
  When tasteful morsels rise!—
As though a greater noise redeems
  The lesser enterprise!

These taints of longshore life betray
  Your source, where song is wrong:
They do and dare in silence, they
  To whom the seas belong.

No, Seamew! for the ocean deep
  You're not full-fitted yet;
But, while those mighty wings you keep,
  You need feel no regret.

Your dainty yacht for pleasure trips
  Will serve in breezes soft;
And, when the storm-fiends come to grips,
  You'll wheel your course aloft.

And mount and glide and swerve and swoop
  And soar upon the gale—
Close-reefed one moment dive and droop,
  Then race away full-sail.

Who is it, Cormorant, now who laughs
  In his empire of the air?
No question now of surplus gaffs,
  Or what to trim and pare!

Go south, and tell your cousin lone,
  Who knows not of his loss:
Here's one who has usurped his throne,—
  Challenged the Albatross!

1920.

Appendix:

A LIST OF WALTER GARSTANG'S OTHER PUBLICATIONS

1. PLYMOUTH (AND MANCHESTER), 1889–1894. (*Marine Biology*)

1889 Report on the Nudibranchiate Mollusca of Plymouth Sound. *Journ. Mar. Biol. Ass., I,* pp. 173–198.

1890 Foreign Substances attached to Crabs. *Nature,* XLI, pp. 417, 490, 538.

,, A complete list of the Opisthobranchiate Mollusca found at Plymouth, with further observations on their Morphology, Colours and Natural History. *Jour. M.B.A.,* I, pp. 399–457.

1891 Report on the Tunicata of Plymouth, I.—*Ibid.,* II, pp. 47–67.

,, Note on a new and primitive type of Compound Ascidian (*Archidistoma aggregatum,* gen. et sp. nov.). *Ann. Mag. Nat. Hist.* (6), VIII, pp. 265–8, and *Zool. Anz.,* XIV, pp. 422–424.

,, On some Ascidians from the Isle of Wight. *Jour. M.B.A.,* II, pp. 119–140.

1892 An Attempt to elucidate the real Structure and Relations of Moss's polystigmatic Appendicularian. *Trans. Liv. Biol. Soc.,* VI, pp. 59–76.

,, On the Development of the Stigmata in Ascidians. *Proc. Roy. Soc.,* LI, 505–513.

,, Note on the Structure and Habits of Jorunna Johnstoni. *Conchologist,* II, pp. 49–52.

,, On some new or rare marine Animals recently discovered on the Coast of Devonshire. *Trans. Devon. Ass.,* XXIV, pp. 377–386.

,, Notes on the marine Invertebrate Fauna of Plymouth for 1892. *Jour. M.B.A.,* II, pp. 333–339.

1893 On the Relations of Hesse's Doto uncinata to the genus Hancockia. *Conchologist,* II, pp. 110–112.

,, On some Bipinnariae from the English Channel. *Quart. Jour. Micr. Sci.,* XXXV, pp. 451–460.

1894 Preliminary Note on a new Theory of the Phylogeny of the Chordata. *Zool. Anz.,* XVII, pp. 122–125.

,, Faunistic Notes at Plymouth during 1893–4, with observations on the Breeding Seasons of Marine Animals, and on the Periodic Changes of the Floating Fauna. *Jour. M.B.A.,* III, pp. 210–235.

II. OXFORD, 1894–1897

(*Animal Morphology and Bionomics*)

1894 On some Modifications of the Tunicate Pharynx induced by the violent ejection of water. *Jour. Jun. Sci. Club, Oxford,* Nov., pp. 267–269.

.. On the Gastropod Colpodaspis pusilla of Michael Sars. *Proc. Zool. Soc.,* pp. 664–669.

1895 Budding in Tunicata. *Sci. Prog.,* III, pp. 43–67.

,, The Chromatophores of Animals. *Ibid,* pp. 104–131.

,, Outlines of a new Classification of Tunicata. *Rep. B.A.,* Ipswich, pp. 718–719.

1896 On Doris maculata, a new sp. of Nudibranchiate Mollusk found at Plymouth. *Jour. M.B.A.,* IV, pp. 167–8.

,, The Morphology of the Mollusca. *Sci. Prog.,* V, pp. 38–59.

,, Contributions to Marine Bionomics, I. The Habits and Respiratory Mechanism of Corystes cassivelaunus. *Jour. M.B.A.,* IV, pp. 223–232.

,, On the Aplacophorous Amphineura of British Seas. *Proc. Malacol. Soc.,* II, Oct., pp. 123–125.

1896 On the Function of certain Diagnostic Characters of Decapod Crustacea. *Rep. B.A.,* Liverpool, pp. 828–830.

1897 Contributions to Marine Bionomics. II. The Function of Anterolateral Denticulations of the Carapace in Sand-burrowing Crabs.
Jour. M.B.A., IV, pp. 396–401.

„ Contributions to Marine Bionomics. III. The Systematic Features, Habits, and Respiratory Phenomena of Portumnus nasutus, Latr. *Ibid.*, pp. 402–407

„ On some Modifications of Structure subservient to Respiration in Decapod Crustacea which burrow in Sand, etc. *Quart. Jour. Micr. Sci.*, XL, pp. 207–228.

„ Physiology and the Ancestry of Vertebrata.
Trans. Jun. Sci. Club, Oxford (*N.S.*), I, pp. 1–6.

III. PLYMOUTH, 1897–1902

(*Fishery Investigations*)

1897 Hjort's Hydrographic-Biological Studies of the Norwegian Fisheries: a Review.
Jour. M.B.A., V, pp. 56–71.

1898 Report on the Surface Drift of the English Channel and Neighbouring Seas during 1897. *Ibid*, pp. 199–231.

„ On the Variation, Races, and Migrations of the Mackerel.
Ibid, pp. 235–295.

1899 On the Plankton and Physical Conditions of the English Channel. *Rep. B.A.* (Dover), 3 pp.

1900 The Impoverishment of the Sea.
Jour. M.B.A., VI, pp. 1–69.

„ Preliminary Experiments on the Rearing of Sea-fish Larvae. *Ibid*, pp. 70–93.

„ Evidence before Select Committee of the House of Commons on the Sea Fisheries Bill.

1900 Evidence before the Interdepartmental Committee on Fishery Statistics (Cd. 1063).

„ The Plague of Octopus on the South Coast and its Effect on the Crab and Lobster Fisheries.
Jour. M.B.A., VI, pp. 260–273.

1900 Gadus esmarkii in Shallow Water; and an albino Hake. *Ibid*, pp. 274–276.

1901 Note on the Programme of the Christiania Conference with respect to Fishery and Hydrographical Investigations in the North Sea. *Report of the British Delegates, etc.* (Cd. 1313), *Appendix*, pp. 46–48 (published 1903).

,, The Exploration of the Sea. *The Times*, July 2nd, 1901.

1902 Evidence before the Board of Trade Committee on Ichthyological Research (Cd. 1312).

,, Scientific Fishery Investigations. *The Times*, April 14th, 1902.

,, The International Investigation of the North Sea. *Proc. Brit. Ass.*, '*Times*' *Report*, Sept. 13th, 1902.

IV. LOWESTOFT, 1902–1907

(*International Investigations*)

1903 Fishery Research. *The Scotsman*, April 9th, 16th and 29th, 1903.

,, Report on Trawling and other Investigations in the South Devon Bays, 1901–2. *Jour. M.B.A.*, VI, pp. 435–527.

1904 Evidence before the House of Lords Committee on the Sea Fisheries Bill (H.L.), March 17th, Q. 2145–2362.

1905 Provisional Report on the Natural History of the Plaice (Committee B). *Rapp. et Proc. Verb. du Conseil International*, III, *Appendix H*, 53 pp.

1905 General Report on the Fishery Investigations. *International Investigations, M.B.A. Report*, I. (for 1902–3), p. 1–11.

,, Report on Experiments with Marked Fish during 1902–3. *Ibid*, pp. 13–43.

Experiments on the Transplantation of Small Plaice to the Dogger Bank. *Ibid*, pp. 45–65.

,, Report on the Trawling Investigations, 1902–3. *Ibid*, pp. 67–197.

1906 The Fisheries of the North Sea, and the Bearings of recent Investigations upon the Problems of Supply. *Jour. Soc. of Arts,* LIV, Feb. 23rd, 1906.

1907 Summary Report on the present state of our knowledge with regard to the Plaice and Plaice-Fisheries (in collaboration with Dr. C. G. Joh. Petersen and Dr. H. M. Kyle). *Rapp. et Proc. Verb.,* VII, 98 pp.

1909 The Distribution of the Plaice in the North Sea, etc., according to Size, Age and Frequency (Committee B). *Rapp. et Proc. Verb.,* XI, pp. 65–133.

V. LEEDS, 1907–1933

(*Nil humanum alienum*)

1912 Report on Experiments with marked Plaice, 1904–5. *IVth Report of North Sea Investigation Committee* (Cd. 6125) (1909), pp. 153–224.

1918 The Leeds War Memorial. *Yorkshire Post,* Dec. 3rd.

,, Nature and Man (Presidential Address to Y.N.U.). *The Naturalist,* March and April, 1919.

1919 The Song of the Tree Pipit. *The Times,* May 8th.

,, The Return to Oxford, a Memorial Lay (*Blackwell*).

,, On Sea Fishery Investigations and the Balance of Life. *Nature,* Sept. 18th, pp. 48–9.

1920 To a Herring Gull (on Time's Changes). *Oxford Magazine,* Feb. 6th.

,, Leeds and the Museum Question. *Yorkshire Post,* April 28th.

,, Three Songs and a Sonnet (The White Rose, The Red Spinx. The Homeless Knights, ' To Absent Comrades ') *Presented to the Students of the University of Leeds,* June, 1920.

1921 Darwin up to Date. *Yorkshire Post,* March 8th, 14th, 19th, 24th, April 2nd, 12th, 14th 16th.

,, Friendship (a sonnet). *Nature,* July 7th.

1922 The Theory of Recapitulation: a critical Re-statement of the Biogenetic Law (read June 2nd, 1921).
Linn. Soc. Jour. Zool., XXXV, pp. 81–101.

,, Songs of the Birds, *John Lane*, The Bodley Head (2nd edition enlarged, 1923).

,, The Student's Opera (*Jackson*, Leeds). *Produced at the University Jubilee on December* 18*th and* 20*th*, 1924.

1924 The Songs of Birds (Abstract of a Lecture to the Science Masters' Association).
School Science Review, No. 23, Feb., 1925.

1925 Wordsworth's Interpretation of Nature (Presidential Address to Leeds Philosophical and Literary Society).
Nature, Special Supplement, Jan. 16th, 1926, and *Proc. Leeds Phil. & Lit. Soc.* (*Literary Section*), I, pp. 72–90.

1926 (*With Sylvia Garstang*). On the development of Botrylloides and the Ancestry of Vertebrates (Preliminary Note).
Proc. Leeds Phil. & Lit. Soc. (*Scient. Sect.*), I, pp. 81–86.

1926 Evidence on behalf of the British Trawlers' Federation at a Conference on March 11th, 1926, with the Fisheries Department of the Ministry of Agriculture and Fisheries *re* proposed Regulation of North Sea Plaice Fishery.
Report, B.T.F., Grimsby, pp. 26–39, 55, 60.

1926 Plaice in the North Sea (Effects of War-time Closure).
The Times, April 21st and 26th.

1928 (*With Margery Platt*). On the Asymmetry and Closure of the Endostyle in Cyclosalpa pinnata.
Proc. Leeds Phil. & Lit. Soc. (*Scient. Sect.*), I, pp. 325–334.

,, (*With Sylvia Garstang*. On the Development of Botrylloides and its Bearings on some morphological Problems.
Quart. Jour. Micr. Sci., LXXII, pp. 1–49.

1928 The Morphology of the Tunicata, and its Bearings on the Phylogeny of the Chordata. *Ibid*, pp. 51–187.

,, The Origin and Evolution of Larval Forms. Presidential Address, Section D. *British Ass.*, Glasgow, 23 pp.

,, The Ballad of the Veliger (*to accompany the preceding*).

1929 On the ' dextricolic ' Condition in Tunicates. *Proc. Leeds Phil. & Lit. Soc.* (*Scient. Sect.*), I, pp. 506–515.

,, Wordsworth's Green Linnet. *Nineteenth Century*, September.

,, Natural Selection (a critique). *Nature*, September 14th, p. 410.

1930 Articles on 'Amphioxus,' ' Fish Culture ' and ' Tunicata ' in *Encycl. Brit.*, XIVth ed.

,, The Buckland Lectures for 1929: (1) Frank Buckland's Life and Work; (2) Fish-Farming, old and new; (3) The Balance of Life in the Sea (*republished by the Fishing News, 20, Broad Street, Aberdeen, as a booklet under the Buckland Trust, price 6d.*).

1931 The Phyletic Classification of Teleostei. *Proc. Leeds Phil. & Lit. Soc.* (*Scient. Sect.*), II, pp. 240–260.

,, Embryology and Evolution (a rejoinder to Prof. MacBride). *Eugenics Review*, XXIII, April.

1933 Report on the Tunicata of the ' Terra Nova ' Expedition, 1910, I. Doliolida. *Brit. Antarctic Exped.*, IV, 6, pp. 195–251

VI. OXFORD, 1934–1948
(*Retirement*)

1935 (*With Elizabeth Georgeson*). Report on the Tunicata Part II, Copelata. Brit. Antarctic (' Terra Nova ') Exped., Zool. IV, 8, pp. 263–282.

,, Bermuda and the Gulf Stream. *Discovery*, December, 1935.

1935 Songs of the Birds, 3rd edition, with additional chapter on the Classification of Birds.
(John Lane's 'Week-end Library', The Bodley Head).

1936 (*With Elizabeth Georgeson*). Report on the Appendicularians of the Nautilus (Arctic) Expedition.
Linn. Soc. Jour. Zool., XXXIX, pp. 411–412.

,, Seasonal Changes in the Underwaters of Bermuda.
Nature, 138, July, pp. 60–63.

1937 The Anatomy and Relations of the Appendicularian Bathochordaeus, based on a new species from Bermuda (B. stygius, sp. n.).
Linn. Soc. Jour., Zool. XL, pp. 283–303.

,, On the Size-Changes of Diatoms and their oceanographic significance.
Jour. Mar. Biol. Ass., XXII, pp. 83–96.

1938 (*With R. Gurney*). The Descent of Crustacea from Trilobites, and their larval relations.
In 'Evolution', Essays for Goodrich, *Clarendon Press*.

1939 Nomenclature of Nudibranchiate Mollusca.
Nature, vol. 144, p. 481.

,, Spolia Bermudiana. 1. On a remarkable new type of Auricularia larva (A. bermudensis, n. sp.). 2. The Ciliary Feeding Mechanism of Tornaria.
Quart. Jour. Micr. Sci., vol. 81, III, pp. 321–365.

1940 Stormer on the Appendages of Trilobites.
Ann. Mag. Nat. Hist. (ii) VI, pp. 59–66.

1946 The Morphology and Relations of the Siphonophora.
Quart. Jour. Micr. Sci., vol. 87, II, pp. 103–193.

1948 On the Tornaria grenacheri of MacBride.
Ann. Mag. Nat. Hist. Ser. II, vol. XIV, p. 878–884.

THE ORIGIN AND EVOLUTION OF LARVAL FORMS

SECTION D.—ZOOLOGY.

THE ORIGIN AND EVOLUTION OF LARVAL FORMS.

ADDRESS BY

PROF. WALTER GARSTANG, M.A., D.Sc.,

PRESIDENT OF THE SECTION.

THE transformations, or metamorphoses, of animals have always provided one of the most fascinating chapters of Descriptive Zoology. Their significance in relation to the doctrine of Evolution was a subject of animated debate by previous generations of zoologists, and figured largely in several Presidential Addresses to Section D, notably in those of the late Prof. Milnes Marshall, in relation to the theory of Recapitulation, at the Leeds Meeting in 1890, and of the late Prof. Miall, from the standpoint of Adaptation, at the Toronto Meeting in 1897. The conclusions arrived at by these two distinguished predecessors of mine were by no means concordant, and I hope I am not wrong in thinking the time ripe for reopening the subject. I propose, however, to take it from a third standpoint, distinct from theirs, yet related, which I may broadly define as the part played by larval forms in the course of evolution.

If we take any large class of marine Invertebrates the members of which can be seen to have made substantial progress along one or more lines of descent, a comparison of their larval forms shows that on the whole a larval evolution has taken place more or less parallel to that of the adult evolution, but subject to conspicuous deviations. Primitive types of larvæ are limited to the lower or more primitive sections of the class, and secondary larval characters become more and more pronounced in the higher and more recent members. In this general statement I am thinking of classes like Mollusca and Crustacea, in which the metamorphosis is gradual and continuous, and is not subject to sudden and radical changes of plan, such as are exhibited for example by Echinoderms and Polyzoa.

In Mollusca the primitive type of larva is obviously a Trochosphere, closely resembling that of Annelids in its pear-shaped body, præoral ciliated ring or prototroch, apical tuft, and absence of special Molluscan features such as shell and foot. It is found in each of the main sub-classes of Mollusca, except the Cephalopoda, viz. in *Chiton* (Amphineura), *Patella* and *Acmæa* (Gastropoda), *Dentalium* (Scaphopoda), and *Nucula* and *Yoldia* (Lamellibranchia or Bivalvia). All these are genera which, either in Mollusca as a whole, or in their respective sub-classes, retain a distinct preponderance of archaic characters—*Patella* and *Acmæa* belonging to the lowest section of Gastropoda (Zygobranchia, in spite of loss of the original gills!), *Nucula* and *Yoldia* to the lowest section of Bivalvia (Protobranchia). But in the course of their career as free larvæ, these Molluscan trochospheres all acquire new and divergent features: the trochosphere of *Chiton* lengthens out and develops a dorsal series of cuticular, partly calcified, plates; that of the Limpet acquires a shell which is successively plate-like, cap-like, and nautiloid, its body undergoes the Gastropod torsion, and it then develops an operculum; the *Dentalium* trochosphere develops a pair of mantle-folds and a saddle-shaped shell, which becomes tubular by ventral concrescence of its edges; the larval *Yoldia* acquires a hinged bivalved shell, and both it, *Dentalium*, and *Patella*, but not *Chiton*, develop a foot.

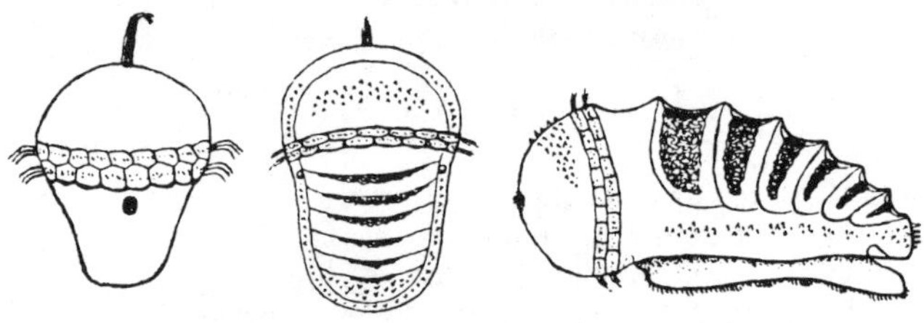

FIG. 1.—Larvæ of Chiton.
A, *C. marginatus*; B, *C. polii*.

It is readily seen that almost all these characters which the trochospheres acquire during their pelagic free-swimming career are in the direct line towards their respective adult characters. As soon as the rudiments of the shell have made their appearance, the larva of *Nucula* is definitely a Bivalve, that of *Dentalium* a præ-Solenoconch, that of *Patella* a Univalve, and that of *Chiton* Polyplacophorous. The secondary characters which appear are essentially adult characters in the making. They have mostly no relation to a pelagic career (*e.g.* the shell-plates of a *Chiton* larva), and may even be an encumbrance—witness the useless digging foot of the *Dentalium* larva—yet they appear. They can also be no heirlooms from pelagic ancestors, since shell and foot speak unequivocally of the ground—the archi-Mollusk was a benthic, not a pelagic animal. These secondary larval characters then are mainly anticipations of adult characters.

But they are not entirely of this nature, for among the examples mentioned the larva of the Limpet develops an operculum which is not present in the adult stage. The early trochospheres of *Dentalium* and *Yoldia* also show features which are both absent in the larva of *Chiton* and have no direct relation to their adult characters. Let us examine these cases a little more closely.

The trochosphere of *Chiton* has a simple prototroch consisting of two parallel rows of cells. As its body elongates the rudiments of six shell-plates arise behind the prototroch, apparently in metameric order from before backwards. During the pelagic career of the larva these plates remain cuticular and uncalcified; but, as growth proceeds and weight increases, the larva swims less and less freely, and takes to gliding along the bottom by means of its pedal cilia. Calcification of the plates then sets in, again in order from before backwards. Simultaneously the creeping sole becomes enlarged by the development of muscles, and these effect attachment above to the developing plates. A cephalic plate in front of the prototroch, and an anal plate behind, are added, thereby completing the typical eight. The prototroch is then absorbed, and the adult life begins. The larval history is thus very similar to that of a simple Polychæte, although segments, in the strict sense of the term, are absent. As in Polychætes, also, the adult characters are not completed until the creature has descended to the bottom.

In the case of *Dentalium*[1] the trochosphere starts with a much more powerful prototroch of three rows of ciliated cells, and goes much further than that of *Chiton* in its development of adult characters during its free-swimming career, for it not only establishes the complete form of its tubular shell—which is much more elaborate than that of *Chiton*—but also develops its characteristic digging foot. There is plainly an adaptive connection between these two features: development of the additional adult characters has been conditioned by the greater ability of the larva

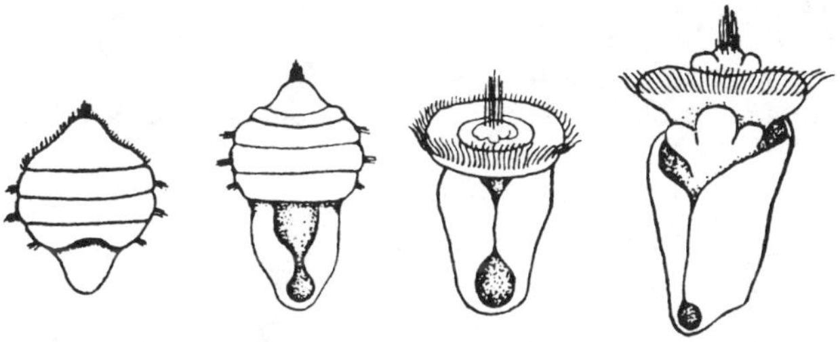

FIG. 2.—Larvæ of Dentalium.

to carry them. The ciliated prototroch is actually extended over part of the surface of the larval body by means of an internal duplicature of the skin behind it, the locomotive girdle projecting freely over the front of the body, like a collar over a coat.

This adaptive modification is carried to an even greater extent in the Protobranch Bivalves *Nucula* and *Yoldia*.[2] The collar becomes a great ciliated cloak or overall, and the duplicature is so deep and precocious that the whole post-trochal region is developed under cover of its five rows of ciliated cells, the middle three of which bear powerful flagella. At the end of the larval period a diminutive adult, fully formed, is deposited on the bottom by disruption of the prototrochal envelope or 'test.' Quick-change artistes are obviously not limited to the human species.

Embryologists are familiar with many other illustrations of this kind of development, *e.g.* the North Sea *Polygordius, Sipunculus*. It shows by easy steps how the more dramatic metamorphosis of *Pilidium* into a

[1] *Dentalium*, Kowalevsky, *Ann. Mus. Hist. Nat.*, *Marseille*, I, 1883. According to the earlier account by Lacaze-Duthiers, the young trochosphere has no less than seven ciliated girdles, four of which give rise to the prototroch by a process of concentration, but their relation to the rows of cells was not described (*Ann. Sci. Nat.* (4) VII, 1857).

[2] *Nucula* and *Yoldia*, Drew, *Q.J. Micr. Sci.*, XLIV, 1901.

Nemertine may have arisen. However, without ranging further afield, these few examples may perhaps suffice to illustrate several important propositions :

(1) the larva has a double task to perform, viz. to distribute the species and to grow up into the adult ;

(2) of these tasks the first is essential, and the second subsidiary—to be undertaken only so far as the larval resources permit ;

(3) the performance of the two tasks together requires the maintenance of an equilibrium between the locomotive efficiency of the larva and the adult weight to be carried ;

(4) the locomotive adaptation of the larva may proceed on new lines, paying no respect to phylogeny, and culminating in some kind of metamorphosis ;

(5) the modification of the larva in this way need not affect the organisation of the adult, since the casting of the most hypertrophied of ciliated girdles involves only slight processes of subsequent repair.

When we pass from the more primitive and ancient groups of Mollusca to the more modern ones, the larva no longer hatches as a simple trochosphere, but is provided with a shell and foot from the first, and the simple girdle of cilia which constituted the prototroch is replaced by a much more powerful organ, the *velum*. This applies to all except the lowest members of the Azygobranch Gastropoda and to all Filibranch

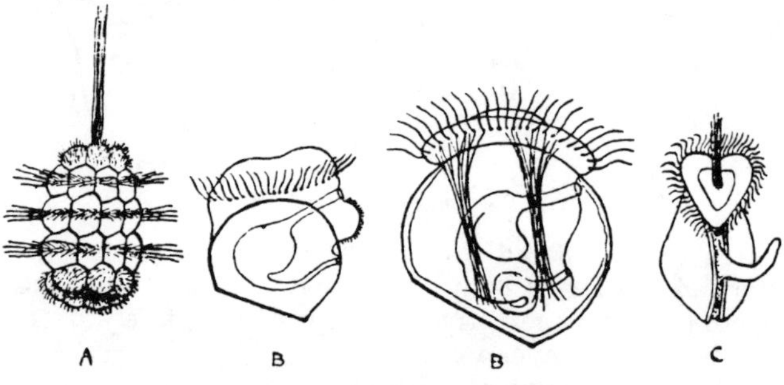

Fig. 3.—Larvæ of Bivalves.

A, *Yoldia* (*Protobranch*) ; B, *Ostrea*, and C, *Dreissensia* (*Eulamellibranchs*).

and Eulamellibranch Bivalves. The velum is only a special development of the prototroch, but by being stretched out at the edge of an extended disk or bi- or tri-lobed frill, the locomotive cilia of the girdle are greatly multiplied in number and power. The larva is the familiar Veliger, though it would be well to restrict this term to the Gastropod larva, and to distinguish the Bivalved form of it by a separate name, *e.g.* Rotiger, from the wheel-like form of its ciliated disk. Both the ciliated arms of the Veliger and the disk of the Rotiger can be protruded freely from the shell and as easily and completely withdrawn inside it.

There are of course many Gastropods and Bivalves in which, even under marine conditions, the free-swimming larval stage has been secondarily reduced in association with a marsupial or incubatory mode of development. Under these conditions the velum or ciliated disk never attains its full size and is often arrested in a very vestigial condition. Finally, the pelagic stage may be suppressed altogether, and the Whelk emerges from the confinement of its brood-chamber as a diminutive adult, ready at once to pursue its definitive career.

The absence of any larval stage throughout the whole class of Cephalopoda is doubtless due to the locomotive agility of the adult which renders a distributive larval phase unnecessary. Although this explanation applies to few other cases of suppression, the fact seems from one standpoint to furnish the climax of the evolutional sequence we have been considering. For the larval phase, like the seed of a plant, is essentially distributive, and in the evolution of Mollusca we have to some extent seen it shift along the steps of the life-history from a very early, simply organised, shell-less stage, the Trochosphere, to an intermediate shell-bearing stage, from this to the highly adapted Veliger or Rotiger, and finally (if we may here include the Cephalopod and the Whelk) to the adult stage itself, the lower stages of development having been successively relegated to the embryonic period. Broadly speaking, this sequence corresponds with an increase in the yolkiness of the eggs, a very simple and widely distributed means of postponing the hatching period to a more advanced stage of development.

It is probably not without significance that this progressive shift corresponds with a time-sequence observable in the order of appearance of the groups concerned, the groups with free Trochospheres, viz. Zygobranch Gastropods and Protobranch Bivalves dating from the Lower or Middle Cambrian, while the groups with Veliger and Rotiger larvæ, the Pectinibranchs, Opisthobranchs, and Eulamellibranchs, appear to be unknown before the late Silurian. A curious exception, urgently calling for further investigation, is the alleged occurrence of Capulids in the Lower Cambrian.

Although, with fuller knowledge of the facts and of the bionomical conditions, it may be possible to explain the cases of reduction or obliteration of the larval stage in terms of adaptation, it seems more probable that there has been a secular change tending to depreciate the value of dispersal as the seas became stocked with an increasing number and variety of specialised inhabitants. When the adults have become highly adapted to the conditions of a particular kind of terrain (*e.g.* rock-life) a prolonged larval life would be of doubtful advantage which regularly carried a large percentage of the larvæ away from the rock zone altogether and landed them in an area of sand and mud.

On the other hand we cannot overlook Prof. Tattersall's *Littorina*,[3] which, not content with all the conventional larval stages, has started a new distributional device of its own by setting adrift the egg-case as well, remarkably adapted to that end.

It is with larval origins, however, not suppressions, that I am now concerned. To some zoologists this question does not arise, or at least presents no serious difficulties. With them larval stages represent foregone ancestors, and all they have to do is to account for discrepancies. As the chain of adult ancestors is drawn out, at each new evolutional

[3] *Littorina*, Tattersall, *Fisheries, Ireland, Sci. Invest.*, 1920, I.

advance the former adult is succeeded by a new one, and slips back into the ontogeny as a developmental stage. Let me briefly state why I am unable any longer to accept this theory. Firstly, it assumes that new steps in evolution are first manifested at the end of the ontogeny, *i.e.* in the ordinary course of adult life. I can find little or no evidence which supports this proposition, and an overwhelming mass of evidence which points against it. An example or two in Mollusca will be brought before you for consideration. Yet this assumption has even been used to support the theory of the inheritance of functional modifications acquired during the active life. Secondly, it is inconsistent with the actual course of development, which often preserves ancestral modes of development of individual organs, but as often as not introduces different organs at periods independent of any probable phyletic time-scale. The totality of an ontogenetic stage is thus normally different from the *tout ensemble* of any ancestor. Thirdly, it ignores what I regard as the chief outcome of modern Genetics. When this subject was last discussed in Section D, Mendel's principles had not been heard of, and Galton's Law of Ancestral Inheritance was the only generalisation in the field. There was nothing then to prevent us from assuming, and much to persuade us, that somehow or other the successive stages of growth were the expression of successive inheritances. To-day, on the other hand, such a phrase seems an anachronism. I feel bound to assume that development is the expression of a single inheritance. I take it that, whatever I may think as to the resemblance between this ontogenetic stage and that extinct ancestor, I may not assume any inheritance of the ancestral stage itself. My boy may be like his maternal great-grandfather and his sister like her paternal grandmother, but, as the phylogeny has been the same, the ancestral stages as such have obviously not been inherited ; and we now know why, or rather how, that comes about.

Viewing development then as the sequential expression of a single inheritance, Science confirms Wordsworth's observation of more than a century ago (1802) that

'The Child is father of the Man,'

and, subject always to the influence of environing conditions, our stages of development are 'bound each to each' by a necessitarian chain of progressive differentiations, each stage depending on its predecessor and determining its successor. The bearings of this doctrine on the problem before us do not appear as yet to have been fully appreciated, but squarely faced, they present issues which are of fundamental importance.

We have seen in the life-histories of *Dentalium* and *Yoldia* that a particular larval organ, the prototroch, can undergo considerable adaptive changes with great advantage to the race, and after serving its purpose can be absorbed, if small, or cast aside, if large, without leaving even a scar. You will note that the unity of the inheritance, and the necessitarian sequence, are not broken by this phenomenon. The prototroch is not a preliminary stage in the formation of any adult organ. If you regard the adult as the final complex resulting from a number of differentiating cell-lineages, the prototroch is only a little subsidiary twig near the base, on which nothing else depends: it can be pruned off without injury to the rest of the series.

But we have also seen that the cell-lineages leading to certain adult organs may differentiate so quickly as to make the rudiments of these organs manifest in the trochosphere of which originally they did not form a part. What will happen if these partly differentiated rudiments should be capable of useful modification subservient to larval as distinct from adult ends? They will, *ex hypothesi*, be subject to the unity of the inheritance, and if the modification be irreversible, *i.e.* incapable of subsequent rectification, the adult form of the same organ will inevitably be affected. Thus some modifications of adult characters may be the result of larval mutations. Is there any evidence that such is ever the case? I believe such evidences are widespread, and that it is only the dominance of an erroneous hypothesis which has prevented us from recognising them before. Let me submit one or two examples in Mollusca for your consideration.

The systematic study of Mollusca has resulted, like that of other groups, in the production of a classification based on the principles of 'adult seriation.' Groups and sub-groups are defined ostensibly by their possession of certain combinations of positive characters; but the real basis is the occurrence of gaps, some large and deep, others slight, in the series of adults available for examination. As knowledge increases, these gaps are often reduced or filled up, and the positive characters defining the groups are then altered accordingly. But some gaps in the seriation remain obdurate: the more we know the sharper they become.

The main lines of Molluscan classification have long reached a stable condition: the gaps between the main sub-classes have undergone no reduction in the time of any of us here, in spite of an immense outpouring of new species and genera, trimmings and rearrangements of families and orders, recent and fossil, and in spite of a considerable increase in our knowledge of their comparative anatomy and embryology. I take the following scheme from Prof. Naef's recent and admirable revision[4] of the Morphology of the group (1926), changing it only by omitting a problematic group of ancient cone-shells (*Hyolithes*, *Conularia*, &c.), usually classed as Pteropoda, but which Prof. Naef raises to the rank of an order and terms Odontomorpha, apparently to suggest a relationship with *Dentalium*. In brackets I have added certain synonyms which may be more familiar than the primary terms actually adopted.

MOLLUSCA.

(*Sub-classes and Orders*)	(*Examples*)
I. Amphineura	
1. Placophora	e.g. *Chiton*
2. Solenogastra	e.g. *Neomenia*
II. Conchifera	
1. Cephalopoda	e.g. *Nautilus*
2. Heteroneura (=Prorhipidoglossomorpha)	
i. Gastropoda	e.g. *Patella*
ii. Scaphopoda (=Solenoconcha)	e.g. *Dentalium*
iii. Bivalvia (=Lamellibranchia)	e.g. *Nucula*

[4] Spengel's *Ergebnisse u. Fortschritte*, III, 1913, and VI,2, 1926.

The gap between Amphineura and Conchifera is absolute: the shell in the former consists of a series of plates (or spicules), in the latter of a single plate (calcified from two lateral centres in Bivalvia). No Amphineuran, living or fossil, approaches the Conchifera by showing an enlargement of one of its plates as the possible predecessor of a single shell, and no Conchiferan, living or fossil, approaches the Amphineura by showing any signs of a duplication or segmentation of its shell into metameric plates.

Similarly in Conchifera, whether the Nautiloid or the conical shell be regarded as primitive, no Cephalopod shows any signs of a lateral torsion approaching the Gastropod twist, and no Gastropod exists with paired gills, auricles and kidneys without also displaying a complete torsion of its mantle-cavity and shell through 180° from back to front. The same peculiarity marks off the Gastropoda absolutely from the Scaphopoda and Bivalvia, although in other respects the morphological agreement between these three orders is extensive and detailed and their relationship must be exceedingly close.

Now let us turn to the larval history of a primitive Gastropod, say *Patella*[5] or *Trochus*,[6] and see how this torsion is accomplished. The

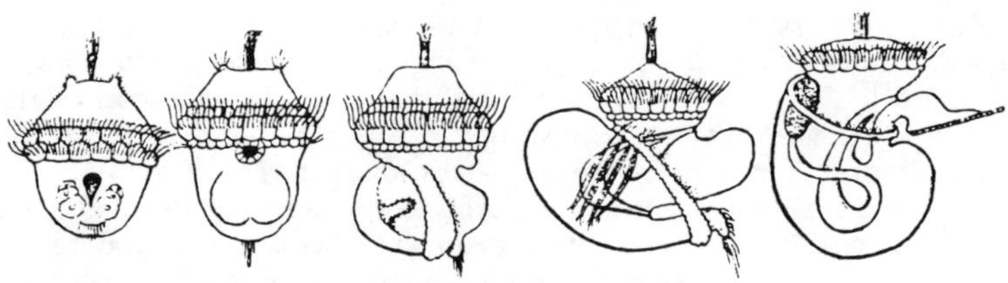

FIG. 4.—Larval Stages of Patella.

trochosphere develops a cap-like shell upon its back, and swims about with it. As the mantle grows more rapidly behind than in front, the additions to the shell are also more extensive behind than in front, so that, relatively to the newer and broader part, the original 'cap' is slowly and steadily pushed upwards and forwards as the apex of a commencing coil. This coil of the larval shell is not quite median in existing forms, but there is reason to believe that it was so in the earliest Gastropods, as shown by the symmetry of the shells in practically all Gastropods known from Cambrian and Ordovician strata (e.g. *Cyrtolites, Sinuites, Salpingostoma, Bellerophon*). Thus the larval shell grows like that of the Pearly *Nautilus* and is at first orientated in the same way: the apex of its coil is directed upwards and forwards over the larval head (exogastric), and a gill-chamber is developed beneath it behind, corresponding to that of *Chiton* and *Nautilus*. At this stage the foot projects freely but carries no operculum; and it is easy to see from the arrangement of parts that an operculum on the foot would be meaningless. For the head is separated from the gill-chamber behind by the whole length of the foot, and, when the body

[5] *Patella*, Patten, *Wien Arbeiten*, VI, 1886; *Acmæa*, Boutan, *Arch. Zool. Exp.*, (3) VII, 1899.

[6] *Trochus*, Robert, *Arch. Zool. Exp.*, (3) X, 1903; and *Zoologie Descriptive*, II, 1900, fig. 508.

contracts, it is the foot, not the head, which can safely withdraw into it, leaving the head, the most vital part of the body, exposed to attack. This vulnerability of the head in the Nautiloid stage of a Veliger is obviously a defect, but it is not long in being remedied. Head and foot as a whole rotate round through 180° until their relations to the mantle-cavity are exactly reversed. According to Boutan, the whole process of torsion is accomplished in *Acmæa* in two or three minutes, so that, as Prof. Naef has pointed out, it is difficult to believe that the change is accomplished by ordinary processes of growth alone. A certain amount of true twisting by muscular contractions would seem to be involved. In *Trochus* (Robert, 1903), the first of all Azygobranchs, the torsion requires six to eight hours. In both forms the shell has already begun its Nautiloid exogastric coil before there is any sign of torsion. In still less primitive forms (e.g. *Paludina*), as Miss Drummond[7] was one of the first to show, the torsion takes longer than in *Trochus*, and starts at a much earlier embryonic stage, before the shell has begun to coil. It is thus probable, as Prof. Naef maintains, that the slow achievement of the torsion by growth-processes spread over a considerable portion of the ontogeny is a secondary modification.

The immediate effect of the change, when completed, is to bring the gill-chamber to the front of the larval body, thus enabling the head, with its all-important velum, to be safely withdrawn into it at the first onset of danger. The foot lastly develops an operculum on its hinder surface, which closes the entrance on contraction.

In the Limpet this rotation is effected during the free larval life, probably as quickly as in *Acmæa*, its next of kin; but in *Trochus* and all subsequent types of Gastropods (Azygobranchia) it takes place in the embryonic phase, so that the Veliger has already undergone torsion before hatching. There can be no two opinions as to the great advance in efficiency shown by the new type of larva as compared with the old. Unfortunately information about the Nautiloid larva of the Limpet and its post-torsional successor is still limited to Patten's observations on specimens reared from artificial impregnations, and neither Patten nor Boutan say much as to the habits of the larvæ. It is also difficult to say whether in its retention of a simple prototroch the larva of the Limpet is primitive or secondarily simplified, but the curious changes and variations which have been described in the structure of its prototroch point rather strongly towards the latter conclusion. The larva of *Acmæa* shows signs of even greater reduction of its prototroch, since the ciliary girdle, though composed of two rows of cells, carries only one row of flagella (Boutan). In *Fissurella*[8] there can be little doubt on this point, for the larva *creeps* out of its egg-shell, instead of swimming, and settles down with the least possible delay to its sedentary rock-life, at once proceeding to absorb the prototroch which it has never used in the open sea, and casting the operculum which it has never used at all. The development of *Pleurotomaria* may some day

[7] *Paludina*, Drummond, *Q.J. Micr. Sci.*, XLVI, 1892; Boutan, *l.c.*, 1899; Naef, 1913, p. 102.
[8] *Fissurella*, Boutan, *Arch. Zool. Exp.*, (2) III, 1886.

reveal the original larval type of a less specialised Zygobranchiate Gastropod.

But the fully developed post-torsional Veliger of an ordinary Azygobranch is thoroughly adapted to an active pelagic career. Its prototroch having now grown out into a pair of velar lobes, the larva no longer rotates like a trochosphere, but directs its movements up, down, or straight ahead on a perfectly even keel. Its velum is so powerful that it can easily sustain the added weight of its partly calcified shell. When suddenly disturbed it reacts in characteristic fashion: its head and velar lobes are immediately withdrawn into the now adjacent gill-cavity, the foot smartly follows suit, and the door is automatically closed by its horny operculum. Owing to its weight the larva falls vertically downwards in the water the moment it stops swimming. As an obscure præ-Georgian poet has somewhere described it:—

'The Veliger's a lively tar, the liveliest afloat;
A whirling wheel on either side propels his little boat;
But when the danger signal warns his bustling submarine,
He stops the engine, shuts the port, and drops below unseen.'

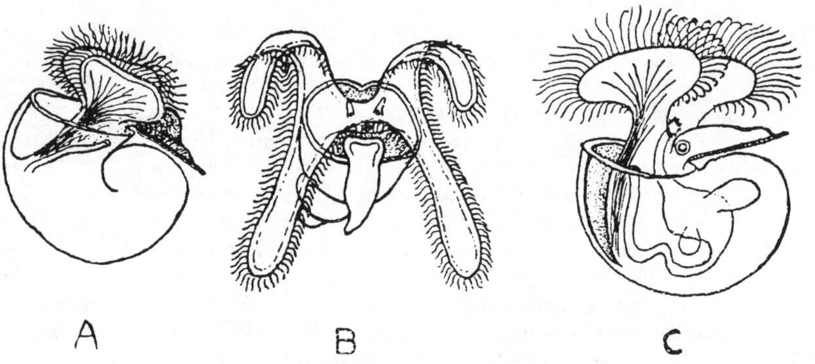

Fig. 5.—Veligers of Azygobranchs.
A, *Nassa*; B, *Dolium*; C, *Opisthobranch*.

Now if I have succeeded in my description of the principal points, I think you will agree that whatever significance may be attached to the twist of its body in an adult Gastropod, there is no doubt about its value to the larva. As to the origin of this torsion, all previous attempts to explain it have been based on the assumption that it arose during the adult life of some early type of Mollusk. I do not propose to go into all these theories, instructive as they are, since their divergences merely illustrate the difficulty of any solution on those lines. Like the asymmetry of *Amphioxus* and the one-sided preponderance of Echinoderms, the torsion has remained a standing puzzle. It has been hardly attempted to assign a utilitarian value to the initial and intermediate stages which must have been required to effect the change in a series of adult ancestors. I will deal later with the further point as to the failure of any of these intermediate links to persist.

In any case I venture to suggest that the torsion of Gastropods arose in the first place very much as you see it develop to-day, as a larval adaptation, in response to larval needs; and that it was perpetuated because, once accomplished, it was of immediate advantage both to larva and adult. It transformed the earlier Nautiloid type of larva into the much more effective Veliger; and the Veliger, settling down to resume the benthic life of its sire, found no serious obstacle to growth in the new arrangement. Lastly, the adult, whose head had previously had little or no protection, was now able to withdraw it on disturbance into complete safety. I assume that the ancestor had a muscular creeping foot, neither so simple as that of *Chiton*, nor so complex as that of a Cephalopod, but definitely suctorial, though capable of progression.

The only mutation required to start the torsion in the assumed ancestral præ-Gastropod larva was an asymmetry in the development of the retractor muscles, thus bending the head and foot round during contraction; but it remains for further investigation to show whether or not the rotation in *Patella* and *Acmæa* is actually determined in this way, as indicated by the rapidity of its accomplishment. The ideal muscular arrangements for bringing about complete rotation would consist of a right-sided cephalic retractor with posterior attachment and a left-sided pedal retractor with an attachment in front of the other, the two crossing one another more or less at right angles. Patten's figures show that these conditions are realised in the præ-torsional stage so far as the right side is concerned (see fig. 4, second figure from the right), but unfortunately leave us in the dark as to the arrangement of muscles on the left side. It is manifest, however, that, owing to the small size of the larval body, any muscular disparity between right and left sides in the direction indicated would conduce towards a reversal of the relations of head and foot to the mantle-cavity at every contraction, while fixation of the organs in the reversed position would be a simple matter at this stage of development, when the body-muscles are just beginning to be actively differentiated, and their connections have yet to be established. Thus, although the theory of a larval origin of the Gastropod torsion cannot be established on our present data, we can at least claim that rotation *may* have been accomplished in the way suggested, and that the ease and rapidity with which it could be achieved contrast favourably with the difficulties besetting any theory of progressive torsion through a long series of adult ancestors. In the larva the smallest twist would produce a favourable change.

In his own ingenious theory of 1913 Prof. Naef has suggested a novel way out of some of these difficulties. He regards the ancestor of Gastropods as a free-swimming Mollusk, not unlike a small *Nautilus* in appearance and habits, but with a more flexible 'neck' or stalk connecting the anterior combination of head and foot (or *Kopffuss*) with the visceral or mantle sac behind; and he associated the origin of Gastropods with a change of habits from swimming to creeping. The Nautiloid position of the shell with the coil forwards and the aperture behind is regarded as convenient for swimming (Prof. Naef does not note that this is only true of *backward* swimming!), but is assumed to have been incompatible with creeping, owing to pressure of the coil on the animal's head and neck. A 'correction' of these arrangements was therefore needed, which has been achieved by the actual torsion. The greatest novelty of Prof. Naef's theory now comes in. He supposes that the earliest Gastropods and their immediate predecessors, owing to the flexibility of their 'necks,' were able to twist their shells round, from back to front, or *vice versa*, at will —as freely, he adds, as a bird can turn its head. On this theory the difficulty as to intermediate stages disappears. A snail that has undergone torsion has not acquired something entirely new: it has merely reversed what may be called the resting attitude of its shell. The power of twisting its neck was not entirely lost until the process of reversal was completed. By that time, however, the snail had given up swimming altogether, no longer needed the *Nautilus* poise, and had settled down to the monotony of a creeping life. As an intermediate stage it is suggested that the symmetrical shells of Cambrian snails (e.g. *Bellerophon*) may have rested sideways, *i.e.* with the spire over the left side of the body and the open end over the right—a position from which it would require, as it were, only half a pull on the left side to bring the shell back to the *Nautilus* position for swimming, or half a pull on the right side to bring the open end forwards into the position most suited to creeping.

For the rest I ought perhaps to add that Prof. Naef explicitly rejects the theory of the Veliger which I have adopted here, viz. that it is a Trochosphere transformed by the incorporation of Molluscan characters, and regards the Veliger as a 'phylogenetic reminiscence' of the pelagic ancestor of Gastropoda, which was adapted for swimming like a Pteropod by means of an expanded and bilobed foot.

You thus have before you two theories in explanation of the same facts, and the only tests by which you can judge between them are the degree to which they conform to well-established facts, and their consistency with the order of events revealed by a wider survey. As I have put before you a rival explanation, I may perhaps point out in what respects Prof. Naef's theory seems to me to be lacking in cogency: (1) We know that the morphological relations, both anatomical and embryological, of Gastropods to Scaphopods and Bivalves are much closer than to Cephalopods, and we are on sure ground when we conclude that the præ-torsional ancestor of Gastropods resembled primitive Scaphopods and Bivalves more closely than it resembled any Cephalopod. This is, in fact, what Prof. Naef's own classification means. By this test two of Prof. Naef's principal assumptions fall to the ground: the adult præ-torsional Gastropod did not possess a narrow flexible 'neck,' or the special muscles required by his theory, or a highly coiled Nautiloid shell. At the point when Gastropods diverged from Scaphopods, and Bivalves, the shell can have been little more than a flat plate. The difference between the two came in with the assumption of a lateral position of the gills in the Scaphopod-Bivalve line and a posterior position in the Gastropod line, thus leading to a preponderating lateral growth of mantle and shell in the former, and a dominating posterior growth in the latter. This entailed an anteriorly directed apex of the shell in præ-torsional Gastropods, as in *Nautilus*, though the resemblance must have been one of simple convergence, since the separation of the whole stock of Prorhipidoglossomorpha from that of Cephalopoda had taken place at an earlier stage, when the shell was presumably still flat. It is apparent from his discussion and diagrams that the tubular shell and elongated body of *Dentalium* have exercised an undue influence upon Prof. Naef's mind as furnishing a kind of connecting link between Gastropods and Cephalopods. They are, of course, indubitably secondary features. Prof. Naef's comparison between the apical slit in the shell of *Dentalium* and the marginal slit of the lower Gastropods will be dealt with at a later stage. (2) Having concluded on these grounds that the common ancestor of Gastropods, Scaphopods and Bivalves possessed a flat shell and no narrow waist capable of rotation between *Kopffuss* and visceral dome, it is easy to see that the phyletic linkage of this group to still lower forms of Mollusca must be with forms of the Placophoran rather than the Cephalopod type. I do not mean that the Conchifera do not form a natural assemblage, but simply that the first Conchifera must have been essentially *Chiton*-like in organisation except for the simplicity of their shell: the conversion of the discoidal shell of the earliest Conchifera into cones, tubes, and spires has taken place independently in Cephalopods, Scaphopods, and Gastropods, in relation to very different habits of life.

If these considerations are well based, there can be no presumption in favour of a pelagic ancestry of Gastropods, and the attempt I have made to explain the evolution of the Veliger larva without regard to such 'phylogenetic reminiscences' can be submitted without anxiety as to objections on that account. The only pelagic feature of a Veliger is its velum, and that, as we have seen, comes down from a Trochosphere, not from a pelagic Mollusk. The bilobed origin of the foot in *Patella* and *Trochus* admits of various alternative explanations.

It will be noted that Prof. Naef's argument in support of the sudden and muscular character of the original process of torsion remains unaffected. He constructed his case from the observation of larval behaviour, but applied it to the behaviour of hypothetical primæval adults, which could not possibly have behaved like larvæ if they had existed. On the other hand, if you prolong backwards into the Cambrian the larval sequence which is demonstrable to-day, and project into it a continuation of the train of modifications in the mode of development of the torsion which we have also seen to be operating, step by step, and sub-order by sub-order, there appears to be neither speculation nor hypothesis in the conclusion that torsion in Gastropods arose as a larval mutation: the logic is that of simple mathematical extrapolation, or of projecting a curve the equation of which is known.

Of course I cannot tell whether you consider my proposition reasonable, or not, on the evidence I have put before you, and I have sought to base it entirely on positive grounds which are open to verification. Let me, however, now draw your attention to the secondary or corroborative evidence. At the outset of my discussion I remarked on the sharpness of the gap which separates Gastropoda from all other groups of Mollusca. The one thing, *i.e.* the only thing of importance, that distinguishes Gastropoda from other Conchifera is their torsion, and that is complete from the start. Torsion makes the Gastropod, and it appears in the systematic sequence as a true saltation. Now if torsion arose in the first instance by gradual modifications of adult form, each step fitted to some particular combination of external conditions or internal functionings, surely somewhere over the wide earth we ought to have found a Zygobranchiate snail with its torsion incomplete. There are, I believe, 180 degrees in a half-circle. Allowing 10 degrees as a reasonable range for each successive stable position in a series of adult modifications, we have eighteen different positions in which some snail or other might reasonably be expected to have made a halt in the orthogenetic advance.

In the Opisthobranchiate Mollusca there is abundant evidence that an evolutional process of *detorsion* has actually occurred. Anus, gill, and kidney, at first in their mantle-chamber, have travelled back again along the right side of the body, reversing the original order of events. (There can be no confusion between the stages of retreat and those of advance because all of these unwinding snails have come back without certain organs of the original left side with which they went forward.) Every possible stage from complete torsion through partial to complete detorsion —far more than the eighteen grades which I assumed—is represented to-day by families, genera, and countless species—*Actæon*, *Bulla*, *Philine*, *Aplysia*, *Eolis*, *Doris*, &c., not to speak of the Pteropods derived from them. Their variety shows us what must have occurred on the forward march of the præ-Gastropods if it proceeded by comparable stages ; and although many that went forward would certainly fall out in the long lapse of time owing to changed conditions, yet there must have been opportunities for adaptations capable of preserving the essentials of one or more of these eighteen advancing types. We know this from facts. In each minor group of primitive snails, possessing clear remains of the original bilateral symmetry of gills, auricles, and kidneys, there are genera which have come down to us unchanged from Silurian times at least, e.g. *Patella* and *Acmæa*, *Pleurotomaria*, *Turbo* and *Trochus* ; and scores of other Zygobranchiate genera exist which only differ from their Cambrian ancestors in trifling details of shell sculpture. Yet not one of these snails falls short of complete torsion through 180°.

It seems impossible to avoid the conclusion that the gap in the adult succession between normal symmetrical Mollusca and Gastropoda is due to some cause other than natural extinction or the imperfection of the 'biological record' to be read in the existing fauna. The gap marks an evolutional saltation. The Gastropod accordingly is a 'sport,' and is the consequence of a sudden jump in the evolution of the Veliger larva in Cambrian, possibly earlier, times. With its visceral dome reversed this new larva settled down and grew to maturity, the general course of growth being unaffected by the change. When its growth finished, the first Gastropod had been created. How far the first Gastropod differed in other respects from its predecessor would require a long argument to tell, except as regards one character to be dealt with in a moment. There could be no trouble over its reproduction, since in *Chiton* and the Zygobranchs eggs and sperms are shed into the sea. The new characters were presumably dominant : the recessives, if any are now left, are apparently non-viable. Whether a gene was added, or dropped, I leave to geneticists.

At this stage I dare say the thought may be crossing the minds of some of you that snails always have been queer-looking things, with something abnormal about them, and that definitely to label them as 'sports' will not seriously disturb any cherished convictions. Even if the abnormality first appeared in a Veliger, that is merely to say that one larva went wrong, whereas most larvæ behave properly. Let the Gastropod go into the same pen with Darwin's Niata cattle, and its veliger with the abnormal embryo that produced the La Plata race—what then? Stands not Scotland where it did?

You will note, however, that Gastropods form no inconsiderable section of Mollusca, and that Mollusca constitute one of the nine large phyla into which the animal kingdom is divided. A few years ago I brought a case very similar to this, but without any touch of abnormality about it, to the notice of the Linnean Society, and claimed that the carapace of Crustacea was also in the first instance a larval adaptation in some primitive Trilobite. If that case holds too, as I firmly believe, and as I hope before long to establish in full detail, the whole phylum of Crustacea must be added to Gastropoda and the Niata cattle. Last year, at Leeds, I put forward some new grounds, now published with fuller details, for the conclusion that Appendicularians are not primitive Tunicates, to be acknowledged by Ascidian tadpoles as their ancestors, but Doliolids gone astray in their development. The larval form in this case has ousted the adult from its supremacy in the life-history and has created a free-swimming pelagic creature out of originally sessile ancestors.

In short the Gastropod and its Veliger loom large in this address, not as ends in themselves, but as an additional example of a wide-ranging phenomenon. The man in the street scoffs at the idleness of the question 'Did the hen come first, or the egg?' He thinks it one of Nature's insoluble mysteries, but admits the priority of the egg when it hatches into a monster with two heads or three legs. In a sense I have looked around for a convenient monster, have found it in the snail, and now seek to show that 'the exception proves the rule.' I stick to snails because we are dealing with a problem which requires a certain amount of concentration, and one point assists another.

We left the ancestral Veliger creeping on its rock and growing up into the first Gastropod, and we are to ask if the new position of its visceral dome, twisted round through half a circle, was not attended by some inconveniences. Before the torsion the gill-chamber lay behind, as in *Chiton* and *Nautilus*. It was a more definite chamber than in *Chiton*, but not so big as in *Nautilus*, and its cavity opened downwards behind the foot, not forwards as in Cephalopods, because its walls were not specialised to propel the animal backwards through the water. This is where bionomics comes in to help morphology. When Prof. Naef treats the ancestor of Gastropods as a kind of *Nautilus*, he is putting the cart before the horse, or, more exactly, the specialised condition before the unspecialised, the higher before the lower. Before the mantle-cavity of *Nautilus* was used as a locomotive organ it must have been what it still is in Gastropods, a simple shelter for the gills, and a passage for the products of anus, kidneys, and gonads. This curious combination of cloaca and respiratory chamber, easily explained by its evolution from the condition seen in *Chiton*, implies arrangements for maintaining a through circulation of water, as well as for preventing contamination of the respiratory water by waste products. In Cephalopods the respiratory current is maintained by muscular pulsations, in Gastropods by ciliated tracts, and there can be no question as to which of these methods is primitive. Not only is the respiratory mechanism of *Nautilus* more advanced than that of any Gastropod, but *its use as a locomotive device is also secondary,* being but a further elaboration of the breathing movements. The inferences we drew from the shells of Mollusca are thus confirmed by a consideration of the gill-chambers. It is not *Nautilus* but *Chiton* that shows us most nearly the form of Gastropod (and indeed Conchiferan) ancestors. In *Chiton* the gill-chamber is hardly established as such: the groove between mantle and foot is open at all points, being merely a little deeper behind than in front. Water flows in at the sides, bathes the gills hanging from the roof, and escapes behind where the anus lies between the symmetrical pores of the kidneys, everywhere overhung by the projecting mantle-frill.

And now suddenly these excellent sanitary arrangements have been turned round from back to front through the efforts of a præ-Veliger to get its head into the hole before its foot (the Lamarckism may be excused!), and by the persistence of the larval adaptation. Before the torsion the gill-chamber opened freely behind; after the torsion the free exit of water and waste products was impeded by the snail's head and neck. There were two possible solutions of this dilemma: either the snail must die, and so wipe out the larval mutation altogether—in which event, had it happened, there would have been no order of Gastropoda to perplex the Zoological student—or a new exit must be provided. The latter alternative was followed. Every member of the most primitive section of Gastropods to-day, *i.e.* every snail which possesses paired gills and auricles (Zygobranchia), has in one form or another a slit or a hole piercing mantle and shell where these overhang the gill-chamber in front. It is not present in the larva; it does not appear until the larva has settled down to its permanent life on the bottom. It is an insignificant notch in the simpler cases, and yet it is to the development of this 'breathing hole' that the survival of the whole order of Gastropoda must be ascribed. Bearing in mind the direction of the ciliary currents in *Chiton*—in at the sides, and out at the middle—we must picture the young snail with these arrangements, but reversed now from back to front, and at the outset of its life on the bottom, possibly with one of Dr. Bidder's Torridonian tides swirling over it. Also we must recognise the effects of continuing growth and differentiation—increase of size and gill-surface, multiplication of muscles between shell above and foot-surface beneath, contractions of these muscles pulling the shell down on to the animal's neck, increasing metabolism and output from rectum and kidneys—all demanding increased ciliary activity, and greater outpouring of waste water beneath the middle point of mantle and shell. If I could start again as an Experimental Biologist I would greatly like to try the effect on a growing epithelium of a continued stream of deoxygenated water charged with a suitable quantity of metabolic waste. Would it or would it not inhibit growth at the point affected? In any case, whatever the chain of cause and effect, that is what happens now in the development of every young Zygobranch at the outset of its adult life. Beginning with the intact edge of the larval mantle and shell, the mantle grows less and less freely at the middle point where the waste water is poured out and grows freely everywhere else, with the resultant formation of this so-called 'marginal slit,' which of course is also manifested by a corresponding gap in the shell. As growth proceeds the viscera behind occupy an increasing proportion of the space below the shell, and the gill-chamber itself shifts forwards, so that continual readjustment of slit to cavity is required. In *Emarginula* the mantle maintains the slit at its edge throughout life. As the mantle extends, the slit extends; but the intact part of the mantle behind also extends, and seals up with a secondary deposit the older parts of the slit in the shell. Thus arises a long seam in the shell, the so-called 'slit-band,' which marks the track along which the slit has travelled. In *Fissurella*, the Key-hole Limpet, a different arrangement prevails: as soon as a slit of sufficient size has been

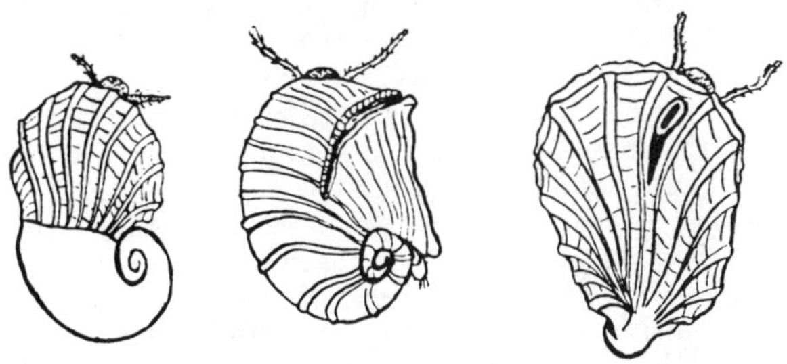

FIG. 6.—Post-larval Development of Fissurella.

produced, the edges of the mantle meet in front of it and fuse, the mantle thus regaining its original integrity. The shell now goes on growing as intact as a Limpet's, leaving a hole near the summit which retains communication with the gill-cavity throughout life, and is enlarged from time to time by absorption. In *Haliotis* the mantle goes on splitting and closing throughout life, thus adding to the number of holes instead of enlarging the first one. The first and its successors are sealed up, one after the other, as the mantle-chamber grows forwards into new positions.

These various arrangements of slit and pores, with numerous intermediate conditions, are adaptive to minor differences of body-form and habits. In some the gills are equal and symmetrical, in others unequal and squeezed to one side; the body may be tall and pyramidal, or flat and broad, the shell accordingly conical or spiral, and carried above the foot or brought down to the substratum. In all, whether symmetrical or asymmetrical, the slit or series of holes lies in the morphological median line, between the two gills, and is associated with a persistence of the original bilateral arrangement of the inhalant respiratory currents. To its presence beyond all doubt the Zygobranchia owe the preservation of their original pair of gills in the reversed mantle-chamber. As soon as the right gill goes (Azygobranchia) the slit goes too, and a new current, oblique in direction, but simple instead of complex, is set up through the chamber, water entering in front on the left side, bathing the persistent (left) gill, then crossing to the right side to which the anus is diverted, alongside the persistent (right) kidney. Entrance and exit are each defined by special folds of the mantle-edge, which may be drawn out into long spouts in front and behind. The evolutional changes in the gill-cavity are somewhat complicated morphologically, but physiologically can be summed up in a single word, sanitation. The various readjustments amount to a series of experiments in the more efficient separation between the respiratory and excretory arrangements, and finally result in the substitution of a simple system which cannot go wrong for one so intricately balanced that it will only work if its owner keeps perfectly still. It is no accident that Zygobranchism is associated with a sedentary rock-life, and that Azygobranchism is distinctive of the snails with versatile and wandering habits.

The true Limpets (Docoglossa) of course gained the same end by different means, sacrificing first one, then both gills in the mantle-chamber, and substituting for them an entirely new system of marginal folds outside the primitive gill-chamber altogether.

From this survey of the facts it seems to be a legitimate inference that the reversal of the mantle-chamber did in fact introduce some serious difficulties into the adult life of the first Gastropods. Retention of the complete ancestral organisation was rendered impossible except by an immediate modification of the mantle margin, and even this permitted no deviation from a very restricted mode of life. 'Radiation' into other environments requiring greater activity was inhibited by the delicacy of the respiratory adjustments, consequent on the partial blocking of the branchio-cloacal aperture. Had the torsion taken place by instalments in successive generations, some of the modifications which were subsequently introduced (with the Azygobranchia) would almost certainly have been accomplished *en route,* and would not have been deferred until the rotation was complete. The nature of the earliest post-torsional modifications thus corroborates the more direct evidence that torsion was, so to say, imposed upon the adult stage, and not primarily developed in its interest.

But the marginal slit has bearings on the general problem which are direct as well as corroborative, since it provides us with a test case of the origin of a typical adult character. We know from Boutan's account of the development of *Fissurella* that there is not a sign of the slit before the sedentary stage is entered upon, and his figures show that an area of shell is produced equal to that of the whole embryonic coil before the marginal slit begins. This area is a mere trifle compared with the ultimate size of the adult shell, but it is enough to show that the slit is a purely adult character and arises at the outset of the adult life.

Now the history of the slit is engraved upon the face of every Zygobranchiate shell, and the date of its commencement in the adult life is to be got by following the 'slit-band' to its source. In every Zygobranchiate living to-day the 'slit-band' begins, like the hole of *Fissurella,* near the apex of the shell in front of the larval coils. Moreover the inscription on the shell is so distinct, and the shell so durable, that it can be read on the shells of the earliest Cambrian and Silurian fossils. Here also in every case, even in the primæval *Bellerophon,* which retains perfect bilateral symmetry in its nautiloid coil, the slit-seam runs up from the margin of the shell nearly to the apex of the coil. There has accordingly been no change from first to last in the period at which the slit develops.

In the first Cambrian Gastropods it must have arisen as a marginal notch almost immediately after the beginning of the adult life, just as it develops

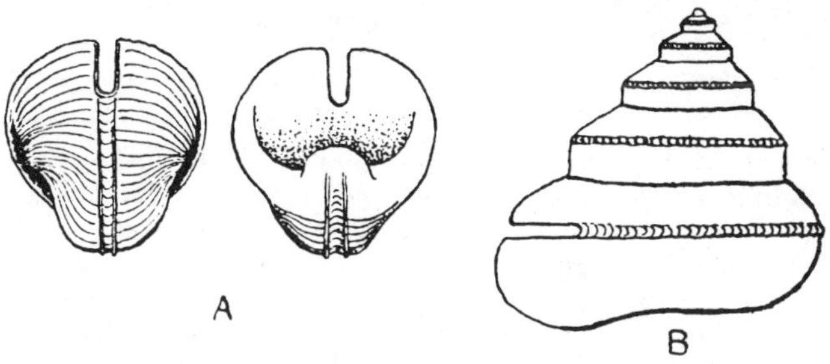

FIG. 7.—Shells of Adult Zygobranchs.
A, *Bellerophon* (*Cambrian*); B, *Pleurotomaria* (*Silurian onwards*).

now; while there is every indication that torsion took place, as it takes place now, during the embryonic or larval, and not the adult, stage.

Prof. Naef has indeed attempted to trace a homology between the apical hole of *Fissurella* and that of *Dentalium*, which, if it could be sustained, would make the slit a præ-torsional instead of a post-torsional modification. He even figures the slit as a feature of the shell before, as well as after, torsion in his diagrams of this process. There is of course a certain correspondence in the position of these two apertures or slits, since both are morphologically median and posterior. But whereas the hole in *Dentalium* is simply a remnant of the original gap between the paired mantle-flaps of the larva, that of *Fissurella* is formed post-torsionally at the extremity of a free median outgrowth of the mantle which has no representative in *Dentalium* or the Bivalves. Moreover, in Gastropods the slit is at right angles to the main mantle-edge: in those Scaphopods which possess a slit as well as a hole, this slit, like the hole itself, is merely a gap between the mantle-folds themselves. These differences are quite sufficient to distinguish the two holes as examples of simple convergence.

Having now put before you all the salient facts as to the history and function of this Zygobranchiate slit, I need scarcely point out to you how admirably these facts serve to disentangle the elements of truth and error which Haeckel so confused in his 'Biogenetic Law.' *Fissurella* has an apical hole which develops by fusion of the lips of a transitory marginal slit. *Emarginula* retains a marginal slit throughout life. It is a statement of simple fact to say in a general way, and with regard to this character, that *Fissurella* goes through an *Emarginula*-stage in its development. But does it follow that the *Emarginula*-stage of *Fissurella* represents an *adult* ancestral condition? On this evidence clearly not, for the adult ancestral condition, *ex hypothesi*, is that of *Emarginula* itself, and *Emarginula* when definitely adult possesses a long 'slit-band' which is completely lacking in *Fissurella*. Let us condense the facts of the two ontogenies symbolically and assume, as is not improbable, that one has been derived from the other. If this assumption is disputed the case for adult recapitulation naturally falls with it. On the Haeckelian hypothesis the inheritance runs like this :—

Emarginula :	Egg—Larva—Adult ↓ (with slit)	
↓		
Fissurella :	Egg—Larva—Neanic stage (with slit)	—New Adult (with hole)

Fissurella is regarded as inheriting all that *Emarginula* has to give, and as then adding a new stage to the series. This differs from the previous adult stage, but is built up out of a Neanic stage, which is claimed to represent the previous adult.

But we have seen that the complete adult ancestral stage is not inherited : the whole of the post-Neanic ontogeny which includes a 'slit-band' is absent from the ontogeny of *Fissurella.* We must therefore distinguish like from unlike, and represent the two ontogenies differently :

Emarginula :	Egg—Larva—	Young Adult— (with slit)	Old Adult (with slit-band)
↓	↓		
Fissurella :	Egg—Larva—	Young Adult— (with slit)	Old Adult (with hole)

From this analysis it results, so far as this one particular character is concerned, that the ontogeny of *Fissurella* repeats the ontogeny of *Emarginula* up to the Neanic stage of the latter, but no further, and then deviates. There is thus no 'compression' of the whole adult stage of *Emarginula* into the Neanic stage of *Fissurella.* It is much the same with regard to all other characters, with this qualification, that the point at which divergence takes place may be quite different for different organs (*e.g.* character of gills, kidneys, &c.). One ontogeny is derivable from another, but when modification is introduced, it is not by the addition of a new total 'stage' at the end of the previous life-history, but by interstitial changes, so to say, either in individual organs or in parts of organs, and usually in quite early stages of growth and differentiation.

I have claimed that torsion arose in the free-swimming larval stage of Gastropod ancestors, and that by so arising it created the order Gastropoda; also that the marginal slit arose in an early post-larval creeping stage to meet respiratory difficulties then first encountered as a result of the mutation. I now claim that the hole of *Fissurella* arose by a modification of the marginal slit at a stage of development scarcely later than that of the slit itself, but at a later period in the phyletic history. By this I mean that the immediate adult ancestor of *Fissurella* was to all intents and purposes an *Emarginula* with marginal slit and long slit-band ; and that the slit was transformed into a hole very much as it is transformed to-day and at the same stage of the life-history.

We have an irresistible tendency when considering the evolution of living things to look for gradual changes—' By Nature's gradual processes be taught ! ' to requote Wordsworth—, but there is no getting over the fact that the conversion of a slit into a hole sooner or later involves an act of discontinuity,—a mutation. At the critical period in the evolutional process one generation had a migrating slit, and the next generation, or some individuals in it, changed the slit into a permanent hole. Now in an organism provided with a free mantle which goes on growing through life, and preserves a slit margin as long as it grows, the mutation to form a hole must be just the same in later as in earlier stages of the migration, and no greater if it occurs at the beginning than if it occurs at the end. We never can be present throughout an act of evolution, for the simple reason that, until countless generations have passed, the mutation is an individual peculiarity, or a local variety, or something not yet sufficiently widespread to ensure our recognition of its significance. Nature's 'gradual processes' of evolution lie not so much in the absence of mutation as in the spreading of a mutation through the community. The nearest approach to witnessing such a process in our time has been the observation by entomologists of the spreading of melanism in moths.

In this address I have sought to keep morphological facts clearly distinguishable from interpretations, but I have also attempted to show that morphological facts require bionomical facts to elucidate their significance. On purely morphological grounds I attempted to show that we are under no intellectual necessity of concluding that everything new must arise late in the life-history, and the development of *Fissurella* shows us that to-day at any rate the marginal slit is converted into a hole at the very outset of the adult life. I am well aware of the fact that there are many other stable conditions of Zygobranchiate holes, and that in *Rimula*, for example, the hole, instead of being apical, is halfway between the apex and the margin. It therefore furnishes to superficial appearances a halfway house between *Emarginula* and *Fissurella*, and renders it perfectly possible, some would say probable, that the immediate ancestor of *Fissurella* was a *Rimula*, with a short slit-band, and not *Emarginula* with a long one. I submit that the existence of *Rimula* makes no difference to the problem of recapitulatory development as evidenced by *Fissurella*. There is a stage in the development of *Fissurella* when its hole is also in the middle, and it is commonly claimed that *Fissurella* on that account goes through a *Rimula* stage after its *Emarginula* stage. But it is equally true of *Rimula* as of *Emarginula*, that it possesses something which *Fissurella* at the corresponding stage does not possess, viz. a 'slit-band,' so that any representation of the definitely adult stage of *Rimula* is absent from the life-history of *Fissurella* as completely as is that of *Emarginula* itself. All that these three genera possess in common is a short transitory post-larval stage with a slit and no band, and it is at this stage that the slit is converted into a hole in *Fissurella*.

Under heredity we cover a multitude of things, and it seems to become increasingly clear that half the things which constantly occur in a given ontogeny, *i.e.* half the links in the necessitarian chain, are not predetermined by intrinsic structure so much as dependent on the operation of influences from surrounding or adjacent parts of the developing organism. At an earlier stage I suggested that the marginal slit itself may have been determined originally—and, I now add, may still be determined—by the pouring out of a horizontal stream of deoxygenated and poisonous water against the growing mantle-edge. Suppose now that in the series of generations between *Emarginula* and *Fissurella* the changing conformation of the body, associated with perpetual downgrowth of the mantle-edge and elevation of the visceral cone, should have gradually involved a relative upward movement in the direction of the exhalant stream. Here without a doubt *Rimula* may find its real significance as a connecting link. The stream is horizontal in *Emarginula*, oblique in *Rimula*, vertical in *Fissurella*. The stream would then continue to play upon one point only, in *Rimula* halfway down the old slit-band, in *Fissurella* at its earliest base. So playing, it would keep the slit open at the same spot throughout and continue to discharge through the same gap. But the mantle, going on with its general growth, would soon prolong the edges of the slit beyond the range of any inhibiting influence from the cloacal stream. The edges would necessarily meet below it, and would there tend to resume their interrupted continuity. The mutation I have spoken of would thereby be accomplished, and its adaptive character would need no separate explanation: adaptation itself would have made the hole, and would have simultaneously ceased to make a migratory slit.

There is an old German proverb which needs to be hung over the mantelpiece of those of us who have a bent for speculation: *Behaupten ist nicht beweisen*. Nevertheless, if art is long, science is much longer, and of all sciences Zoology makes the greatest drafts on time for securing synthetic results. By ourselves in this field we can do nothing. We must critically assimilate the work of our predecessors and co-operate whole-heartedly with our colleagues, or we plough the sands. I trust that I have not misused the presidential opportunity and privilege by this mingled play of criticism and suggestion, and that it may help to clarify some of our evolutional problems. Even if every conclusion it expresses should turn out to be untenable, there are times when it is useful to throw the windows of the mind wide open.